MW00810011

DISASTERS IN SPACE
Tragic Stories from
the US–Soviet Space Race

Hermann Woydt

4880 Lower Valley Road • Atglen, PA 19310

Copyright © 2018 by Schiffer Publishing Ltd.

Originally published as *SOS im Weltraum* by Motorbuch Verlag, Stuttgart, Germany

© 2017 Motorbuch Verlag

Translated from the German by David Johnston

Library of Congress Control Number: 2018937032

All rights reserved. No part of this work may be reproduced or used in any form or by any means—graphic, electronic, or mechanical, including photocopying or information storage and retrieval systems—without written permission from the publisher.

The scanning, uploading, and distribution of this book or any part thereof via the Internet or any other means without the permission of the publisher is illegal and punishable by law. Please purchase only authorized editions and do not participate in or encourage the electronic piracy of copyrighted materials.

"Schiffer," "Schiffer Publishing, Ltd.," and the pen and inkwell logo are registered trademarks of Schiffer Publishing, Ltd.

Cover design by Molly Shields

Type set in Univers

ISBN: 978-0-7643-5617-9

Printed in China

Published by Schiffer Publishing, Ltd.

4880 Lower Valley Road

Atglen, PA 19310

Phone: (610) 593-1777; Fax: (610) 593-2002

E-mail: Info@schifferbooks.com

Web: www.schifferbooks.com

For our complete selection of fine books on this and related subjects, please visit our website at www.schifferbooks.com. You may also write for a free catalog.

Schiffer Publishing's titles are available at special discounts for bulk purchases for sales promotions or premiums. Special editions, including personalized covers, corporate imprints, and excerpts, can be created in large quantities for special needs. For more information, contact the publisher.

We are always looking for people to write books on new and related subjects. If you have an idea for a book, please contact us at proposals@schifferbooks.com.

CONTENTS

The Space Mirror Memorial on the grounds of Kennedy Space Center is made of black granite in which the names of the astronauts killed during American spaceflights are engraved.

INTRODUCTION

Man has been in space for more than half a century, and more than 500 people have been there. Some have been to space more than once. And all who were there describe their trips as unforgettable and incomparable experiences, and it is likely that every single one of them would go again if given the opportunity.

These trips into space involve a number of serious risks. During launch the space travelers are sitting on flying bombs, rockets filled with many tons of highly explosive fuels. The failure of a single component is enough to cause the flight to end before it has begun, in an enormous fireball. In space, space travelers move in fragile vehicles, with just the vessel's thin outer hull to protect them from the most inhospitable of all environments. During activities outside the spacecraft the situation is even more extreme. Only a thin spacesuit protects its wearer from the vacuum, the deep cold, the enormous heat, and micrometeorites. If the space travelers survive the dangers of the flight, they face one last great challenge: the return to earth. During the spacecraft's return to earth, the smallest failure can also lead to catastrophe. Its human occupants must be shielded from temperatures of several thousand degrees caused by friction during reentry into the earth's atmosphere. One or more braking parachutes must slow the vehicle sufficiently to make the impact with the earth or the surface of the water bearable for the occupants.

In the history of manned spaceflight, the launch, travel in space, extravehicular activities, and return to earth have usually been completed successfully. But not always. Booster rockets have exploded, space capsules have leaked, and space vehicles have been destroyed while returning to earth. A number of space travelers have paid for their courage with their lives in these accidents. Several have been more fortunate and survived critical situations.

The American astronaut Virgil "Gus" Grissom, who later died in a fire during a launch simulation, once said in an interview: "If we die we want people to accept it. We are in a risky business, and we hope that if anything happens to us, it will not delay the program. The conquest of space is worth the risk of life."

Although Grissom himself became a victim of space travel, he was nevertheless right. Despite all the setbacks, people are still traveling in space. At some time, however, something unforeseen will happen, perhaps again with catastrophic consequences for life and limb of the space travelers. Let us not deceive ourselves: spaceflight was always dangerous, it is dangerous, and it will always remain so. Let us be glad that there will always be people who are ready to take the risk to expand man's boundaries in science and technology.

Astronaut Dave Scott, one of just twelve people who have had the privilege of setting foot on earth's satellite, made an extraordinary statement while at the Hadley Rille during a moonwalk: "As I stand out here in the wonders of the unknown at Hadley, I sort of realize there's a fundamental truth to our nature. Man must explore. And this is exploration at—at its greatest."

This sentence may explain why the events described in the following chapters do not deter people from going into space. We hope that it remains so.

Monument to the Conquerors of Space in Moscow.

FIRE ONBOARD: APOLLO 1

The race between the United States and the Soviet Union to be first to the moon reached its peak in the mid-1960s. In May 1961, President John F. Kennedy, in a famous speech, had committed the American people to landing a man on the moon and bringing him safely back again by the end of the decade. After that the Americans had taken many small steps to prepare the way for the lunar mission. First, the Mercury Program examined whether and how man could live and work in space. In the following Gemini Program, NASA had tested techniques and procedures necessary for the flight to the moon. Long-duration missions, docking maneuvers, and spacewalks of growing complexity were carried out. The next task was the testing of the lunar spacecraft, the first manned flight of the new three-seat Apollo capsule.

The hurried development of the Apollo spacecraft had not been without problems. The environmental control system had caused problems for North American Aviation, the prime contractor. There had been more or less serious malfunctions during the first unmanned Apollo test flight. For example, there was a short circuit that caused the onboard measuring devices to fail. Much more serious, however, was that the ship's engine had worked properly for only eighty seconds because of a helium leak. As well, during reentry into the earth's atmosphere, a wiring error rendered the capsule uncontrollable.

The failures were analyzed and rectified. Half a year later the second test flight was largely satisfactory. NASA declared the spacecraft operationally ready.

Air Force pilot Virgil "Gus" Grissom was named mission commander. The forty-year-old had flown combat in the Korean War and had been a flight instructor before becoming a test pilot. Grissom, married and the father of two children, was a member of the first group of astronauts, who had been dubbed the "Mercury Seven." In 1961, he made the second American ballistic spaceflight. He just barely escaped death during the landing after the fifteen-minute flight, when his space capsule filled with water and sank after the hatch blew prematurely. He was pulled from the waves at the last minute. Gus Grissom subsequently commanded the maiden flight by the Gemini two-man spacecraft. In Gemini 3, he and Pilot John Young circled the earth three times in five hours. Because of this experience as a pilot and astronaut, Grissom was ultimately selected to command the three-man Apollo capsule.

His crew included thirty-six-year-old Edward "Ed" White, also married and the father of two children. In the 1950s, White had been stationed at Bad Tölz in Bavaria, and in 1962 he was selected to be one of the second group of astronauts. He first went into space in June 1965, with commander James McDivitt in Gemini 4. During this mission, White left the capsule and became the first American to float freely in space, tethered to the spacecraft by just a thin umbilical line. McDivitt and White spent more than four days in earth orbit.

The third member of the Apollo 1 crew was US Marine Corps aviator Roger Chaffee. Like his two crewmates, he was also married with two children. At thirty-two he was the youngest member of the crew. Apollo 1 was to be his first space mission.

Because of the ongoing changes in the design of the capsule, crew training for the first

Above: The crew of Apollo 1 (from left to right): Gus Grissom, Ed White, and Roger Chaffee. Right: The lemon that Gus Grissom hung from the simulator by a wire hanger for all to see. *NASA*

flight had fallen further and further behind schedule. Because of the constant modifications, which were supposed to bring it up to the latest development standard, the spaceship simulator was twice out of operation. The three astronauts had to familiarize themselves with new details and technical changes almost daily and were anything but happy about the situation. Grissom demonstrated his displeasure more than clearly by hanging a lemon in the simulator for all to see.

The continuing delays forced a postponement of the flight, with the internal NASA designation AS 204, which had originally been planned for the last quarter of 1966. Finally, it was scheduled for February of the following year. On January 6, 1967, the spacecraft was finally mounted on the Saturn IB booster rocket, and the combination was transported to the launchpad.

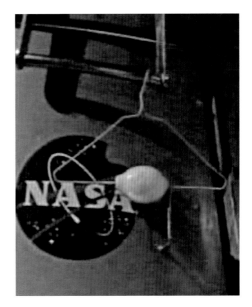

During subsequent preparations, a routine test was scheduled for January 27, one of many such tests. In this simulation, called a

The Apollo capsule is mounted on the Saturn IB booster rocket. *NASA*

Plugs Out Test, all electrical connections between the spacecraft and the launch tower's power supply system were severed. The Apollo spacecraft's fuel tanks were not filled, however,

nor were its pyrotechnics armed. For this reason the simulation was rated nondangerous.

The purpose of the test was to check out the final phase of the countdown, the launch, the ascent, and the first minutes in earth orbit. It had been carried out many times before.

On that January 27, a Friday, the crew was driven to the launch site. The elevator took them up the launch tower, and at about 1300 Florida local time they entered the so-called White Room, 220 feet above the ground. The astronauts had to pass through this room, which could be swung to and away from the capsule on a movable arm of the tower, in order to climb into their spacecraft through the narrow entry hatch. A support team waited in the White Room. Their task was to help the astronauts in their bulky spacesuits climb in and out. Then they helped strap the men to their contour couches and finally locked the hatch. The White Room remained in position for as long as possible, to help the crew get out as quickly as possible in the event of a critical situation. It was

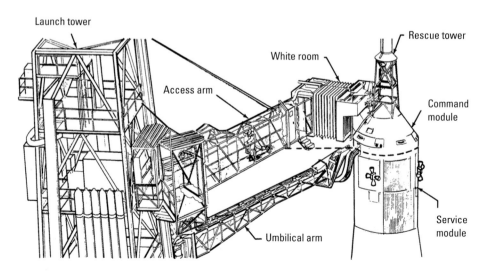

The service structure with movable bridges and the White Room. *NASA*

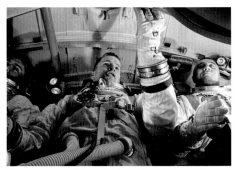

The crew at work in a simulator. From left to right: Roger Chaffee, Ed White, and Gus Grissom.

not pivoted away from the spacecraft until a few moments before launch.

On this day the head of the support team in the White Room, the so-called pad leader, was Donald "Don" Babbitt. With him were two technicians. All three were employees of North American Aviation, maker of the Apollo spacecraft.

The three men helped the astronauts take their places: Chaffee on the couch on the right as seen from the hatch, commander Grissom on the left couch, and White in the center. All three wore sealed spacesuits and were supplied with pure oxygen from the capsule via supply lines. They were hooked up with biomedical sensors. Communication with the launch control room, the so-called Blockhouse, was by radio.

The men of the support team first sealed the inner part of the pressure cabin's three-part hatch. Then they sealed the outer part. Finally, they also applied a seal. After everything was checked to ensure that it was tight, the crew cabin was filled with pure oxygen.

The procedure was carried out following the familiar routine. The only thing out of the ordinary was a strange smell that Grissom had detected when his spacesuit was connected to

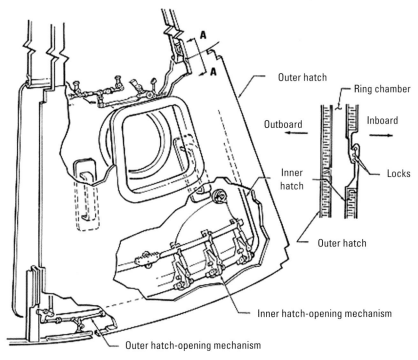

Design of the entry hatch. *NASA*

Night has fallen on Cape Canaveral. Illuminated by searchlights, the Saturn IB booster rocket sits on the launchpad.

the capsule supply system. He said that the smell reminded him of buttermilk. Because of this observation, the countdown was delayed for more than an hour, and an air sample was taken. Whether Grissom's finding was in any way connected to the catastrophic events that followed cannot be determined with certainty, but it seems unlikely.

While the crew worked their way through checklists and the moment of the simulated launch drew ever nearer, increasing communications problems came to light. Contact between the spacecraft and launch control was distorted and the radio messages from the crew could hardly be understood. The countdown had to be stopped several times to allow the technicians to work on the problem. At some point Grissom became annoyed: "How do you expect to get us to the moon if you people can't even hook us up with a ground station?," he growled indignantly.

At about 1800 Florida local time, it was dark at Cape Kennedy on that winter evening. The test sequence, which should have been long over, had not even begun. The switch to the Apollo's internal supply was about to take place. Apart from communications breaks, at that time there were no further anomalies. Nothing pointed to the imminent catastrophe.

The tragedy began at 1830:54.8. Measuring equipment registered a brief voltage increase in one of the capsule's main power lines. From that instant, developments came fast.

Oxygen consumption inside the spacecraft rose steeply until it exceeded the instruments' indicating limits. This was followed by hectic movements by all three astronauts. Afterward it was impossible to determine who moved how. At 1831:04.7 came the first report by radio from inside the capsule, probably from Grissom: "Fire!" Two seconds later Chaffee screamed: "We have a fire in the cockpit!"

The last words from the spacecraft were heard seven seconds later. The transmission was distorted and drowned out by noise. Even after repeatedly listening to the recordings, various witnesses were unable to agree on what the radio message said. It sounded like: "They're fighting a bad fire—let's get out. Open 'er up." Or "We've got a bad fire—let's get out. We're burning up!" Or "I'm reporting a bad fire. I'm getting out!" The last radio message, probably from Chaffee, ended with a scream of pain.

Pad leader Babbitt, who heard the transmissions, screamed to one of his technicians: "Get them out!" Seconds later a jet of flame shot from the spacecraft toward the men in the White Room. They heard a very loud hissing noise. A pressure wave knocked them back.

Sequence of Events

Time (Local Time Cape Kennedy)	
13:00	Crew arrives at the launch site and begins entering the capsule.
13:20	Commander Grissom notices a strange smell, countdown is halted, and an air sample is taken.
14:42	Entry hatch closed; countdown resumed.
14:45	Capsule atmosphere is replaced by pure oxygen.
17:40	Countdown is halted to deal with communications problems.
18:20	Countdown is resumed; communications problems continue.
18:21:11	Readings indicate a rise in cabin pressure.
18:30	Over the radio, noises are heard that can be attributed to movements by Commander Grissom.
18:30:21	Slight increase in White's pulse and breathing rates.
18:30:24	Data from the Apollo capsule's control and navigation system indicate nonspecifiable movements by the crew; slight increase in oxygen consumption by the crew.
18:30:30	EKG readings from White indicate movements by the astronauts.
18:30:39	EKG readings show increased activity by White, but no hectic pace. Control and navigation system indicates an increase in movements by the crew.
18:30:44	Movements by the astronauts end.
18:30:50	Readings indicate a change in the composition of the cabin atmosphere.
18:30:54.8	Clear voltage spike in the capsule's power supply system.
18:30:59	Oxygen inflow rises to above the measured value limits of the instruments.
18:31:00	Movements inside the spacecraft.
18:31:12	Flames become visible to the crew.
18:31:14.7	Astronauts raise the alarm from the capsule.
18:31:16	Cabin pressure rises above the measured value limits of the instruments, and a crack appears in the capsule's outer hull, through which hot gases escape into the White Room, causing major damage.
18:31:16.8	Final radio message by the astronauts begins.
18:31:19.4	Outer hull of the Apollo capsule finally bursts.
18:31:21.8	Last radio message from the astronauts ends in scream of pain.
18:31:22.4	Data transmission from the capsule ends. Monitors show flames and smoke inside the spacecraft.
18:31:25	Fire reaches maximum intensity. Cabin pressure falls to the level of external pressure because of breach in the hull.
18:31:30	Composition of the cabin atmosphere probably lethal. Fire goes out due to lack of oxygen.
18:32:00	Firefighters are dispatched.
18:32:04	Attempts to open the hatch from the outside.
18:32:34	Pad leader reports that the support team has begun rescue attempts.
18:36:31	Pad leader reports that all three parts of the entry hatch have been opened. Firefighters arrive at the accident scene; because of thick smoke the bodies of the astronauts are hard to make out. Attempt to remove White's body from the capsule fails.
18:43	Physicians arrive in the White Room.
19:30	Evidence photos are taken. Recovery of the bodies of the crew begins.
02:00	The bodies of the astronauts are recovered seven and a half hours after the accident.

The capsule had burst under the enormous interior pressure, spewing out thick black smoke and poisonous gases.

The three men had no choice but to flee the White Room and seek safety in the communications arm. Seconds later they had recovered from their initial fright and rushed back. In the dense smoke they looked for fire extinguishers. But there were none. While there were gas masks, they were difficult to put on and then still did not function properly.

Coughing and gasping, the three technicians repeatedly had to flee into the open to catch a few breaths. Meanwhile, on observation cameras, bright lights, smoke, and open flames could be seen through the window of the spacecraft.

Three more technicians at the base of the launch tower had been alerted by screams and joined the desperate struggle to save the lives of the astronauts. They brought a fire extinguisher, but the heavy smoke and enormous heat forced them to repeatedly interrupt their rescue efforts; visibility was zero. The men carefully felt their way closer, but one was at the point of losing consciousness, Babbitt ordered him to get to safety. The man obeyed but returned a few seconds later to lend further help. The technicians in the White Room feared that the heat and flames might ignite the solid-fuel rockets in the rescue tower atop the Apollo. This would surely have meant their deaths.

After some time the heat and smoke finally subsided. The men in the White Room continued their desperate efforts to open the spacecraft's hatch. Finally they succeeded in removing the cover and opening the outer part of the hatch. Their initial efforts to open the inner hatch were unsuccessful, however. They were able to open it a crack, but something inside was blocking it. The interior was so hot that they could not touch it with bare hands.

At 1836:31, Don Babbitt finally reported that the hatch was open. Five minutes and twenty-seven seconds had passed since the crew had first raised the alarm. Four minutes later, firefighters arrived at the accident scene and shined flashlights into the interior of the capsule. Thick gray smoke still filled the capsule, and everything was covered with soot. A jumble of cables and melted plastic covered the floor. There were no signs of open fire.

After the smoke had cleared somewhat, the firefighters were finally able to recognize the bodies of the astronauts, who showed no signs of life. Commander Grissom was lying stretched out on the floor, with his helmet closed, his safety belt opened, and the oxygen lines of his spacesuit still attached to the spacecraft fittings. White lay crossways in the capsule, his helmet also closed, with the partially burned harness unopened. The oxygen line to his suit had been disconnected, the outgoing airline still attached. Chaffee had not left his position on the right of the three couches. His helmet was closed, and the oxygen lines were intact. He had released his safety harness.

The firefighters sent Babbitt and his people to the ground to be treated for their injuries and smoke inhalation suffered during their rescue attempts. At the base of the tower they were met by medics and doctors rushing to the scene. The pad leader informed them that none of the astronauts had survived the fire. When the doctors arrived in the White Room a short time later, they could only confirm his statement. The three astronauts were officially declared dead at 1843 local time.

Efforts to recover the bodies of the dead proved extremely difficult. The enormous heat produced by the fire had welded together melted synthetic materials from the spacesuits and plastic materials from the interior of the capsule. The situation inside the capsule was documented photographically. It took almost seven hours for the firefighters and the doctors to finally remove the bodies.

America's space program had lost its innocence. Everyone had known that there would be victims sometime, but everyone had hoped that it would not happen so soon. And now it had happened. And the way it happened shocked many even more. The victims had not died in the fireball of an exploding rocket, they had not burned up in the atmosphere during reentry, and they had not crashed while attempting to land on the moon. No, they had died in a fire, as it were at their own front door, surrounded by people who tried to help. It was small consolation to know that the Apollo capsule would not forever drift through space like a memorial with the dead astronauts onboard. At least the astronauts could be given a proper burial.

Several weeks before the accident, Grissom had said in an interview: "If we die we want people to accept it. We are in a risky business, and we hope that if anything happens to us, it will not delay the program. The conquest of space is worth the risk of life."

The burnt-out Apollo spacecraft was removed from the booster rocket and taken to an assembly hall for investigation into the cause. Everything that might have had anything to do with the accident was secured. Technicians disassembled the capsule literally to its component pieces. Each individual part was described and its location was documented.

Examination of the bodies of the crew revealed that they had been killed by lethal gases. The astronauts had sustained second- and third-degree burns; however, these were not serious enough to have caused their deaths.

On January 30, a funeral was held at the Kennedy Space Center, and afterward the bodies of the three astronauts were turned over to the families for burial. At the request of the widows of the three astronauts, the mission was renamed from AS 204 to Apollo 1.

The Apollo 1 command capsule after the devastating fire. *NASA*

Gus Grissom was buried in Arlington National Cemetery. Roger Chaffee was buried nearby. Ed White was buried in the cemetery of the military academy at West Point.

The command capsule and the items of evidence that had been removed from it were initially stored in a hangar at the Kennedy Space Center, and one year later they were transferred to a climate-controlled warehouse at the NASA Langley Research Center in Hampton, Virginia. They remain preserved there to this day.

The Apollo capsule was literally disassembled into its individual parts. *NASA*

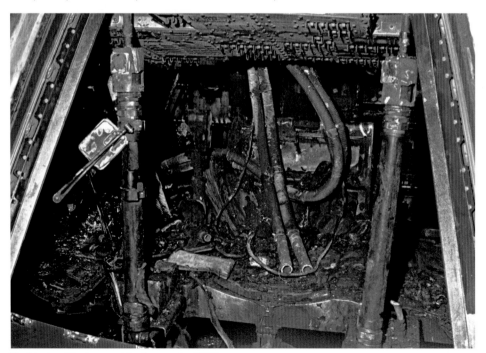

The interior of the burnt-out Apollo capsule. *NASA*

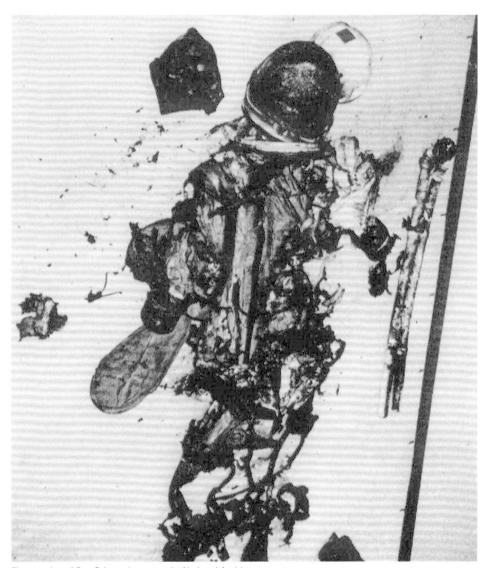

The remains of Gus Grissom's spacesuit. *National Archives*

A hastily convened investigating committee was tasked with determining the cause of the disaster and working out suggestions to prevent a repetition of the accident.

Everything pointed to the fire originating in or near a wiring harness that led along the left side of the cabin at the foot of Grissom's couch,

right in Chaffee's field of view. It was highly probable that crew had been unable to see the flames in the first seconds after they broke out. Six to seven seconds probably passed before the astronauts noticed them.

In the space center in Houston, an attempt was made to replicate the incident in an identi-

Parts recovered from the Apollo capsule were cataloged and examined in a hangar. *NASA*

cal Apollo capsule. Every detail was painstakingly replicated. Finally, in this way the sequence of events was re-created accurately to within the second.

The investigating committee collected its findings in a report published on April 5, 1967. It found that the fire had undoubtedly been caused by faulty insulation on a cable. Six reasons why such a small cause could result in such a catastrophe were listed. The presence of numerous flammable materials in the interior of the capsule, particularly the pure oxygen atmosphere, was viewed as an extraordinary potential hazard. These had promoted the explosion-like spread of the flames. Live wires and conduits with flammable and corrosive fluids were not sufficiently protected against the danger of fire. In addition, there was inade-

quate firefighting and rescue equipment outside the capsule.

The complicated design of the spaceship hatch was blamed for the failure of rescue attempts by the support team. Consisting of three separate parts, the hatch could be opened and closed from the inside or outside by using special tools. Unlike the two outer parts, the inner part could be opened only inward against the pressure of the capsule atmosphere. In the best-case scenario, it took ninety seconds to reach the crew—much too long in an emergency. The hatch was subsequently replaced with an entirely new design. The new one-piece hatch could be operated from inside or outside by means of simple handgrips. In an emergency, the emergency exit could open within two to three seconds.

Many more changes were made following the accident. Effective immediately, spacesuits and the capsule interior were made only of non-flammable materials. Exposed power and supply lines in the spacecraft were guarded against inadvertent damage by the crew. From then on, during the launch phase the astronauts were surrounded by a much less flammable atmosphere, a mixture of oxygen and nitrogen.

The White Room was equipped with an effective ventilation system, and effective fire-extinguishing equipment was installed. Last, a sort of cableway, with which the ship's crew and its support team could slide out of the immediate danger area and down to a bunker on the ground in a matter of seconds, was designed.

These changes took time, however. It would be more than a year and a half before the first manned test flight of what was essentially a new spacecraft, the Apollo 7 mission, took place, in October 1968.

End of main line A wiring harness.

End of main line B wiring harness.

The place inside the Apollo 1 capsule where the fire probably started. *NASA*

DEATH IN A FIREBALL: SOYUZ 1

After the Soviet Union initially took the lead, as the years passed the struggle known as the space race was waged bitterly by the United States and the Soviet Union. The page finally turned and the United States gained the upper hand. They did this even though the Soviets had achieved several spectacular firsts in manned spaceflight in the first half of the 1960s. After the first manned earth orbit by Yuri Gagarin in April 1961, in August of the following year, two Vostok capsules completed the first formation flight by two manned spacecraft. One year later Valentina Tereshkova became the first woman in space. In October 1964, the Soviets put the first multiseat spacecraft into space, and in March 1965, Russian cosmonaut Alexei Leonov became the first man to leave his spacecraft and float in space while tethered to his ship.

But the United States then worked its way forward in small steps with the successful Gemini Program, developing the techniques and procedures necessary to reach and land on the moon. The outstanding maneuverability of the Gemini capsule proved especially helpful, the Russian Vostok and Voshkod spacecraft having no ability to actively change their orbits.

Those responsible for the Russian space program were now under terrific political pressure. The Soviet Union absolutely had to regain the ground it had lost. The first manned landing on the moon was a compulsory social task, since in the Soviet Union, more so than in the United States, victory in the race to the moon would be viewed as a demonstration of the superiority of the nation's political system.

The Soviet Union had not flown a space mission for more than two years. It was time to return to space, and quickly. In the process, the well-being of the cosmonauts was not absolutely treated with the highest priority.

The new Soyuz spacecraft was supposed to help the Soviets take a great step forward. Development had begun in 1963, but it went anything but smoothly. Under the cover designation Cosmos 133, in 1966, the first test failed when the spacecraft could not be placed in a stable attitude in orbit. A planned docking maneuver was canceled and the spacecraft that was to have acted as docking partner was not launched. During attempts to return Cosmos 133 to earth, the braking rockets failed to ignite several times. When they finally did work, the spacecraft threatened to come down in Chinese territory. Provisions had been made for the vessel to self-destruct in such an eventuality, and since no remains of Cosmos 133 were ever found, that at least seems to have worked.

In the second attempt the Soyuz spacecraft did not even make it to orbit. On December 14, 1966, several of the booster rocket's engines failed during the launch sequence. The attempt was aborted on the launchpad. While the fuel tanks were being emptied, the rescue tower atop the Soyuz suddenly ignited, resulting in a huge explosion. At least one member of the ground crew was killed.

The third test, designated Cosmos 140, launched successfully, but then various systems in the spacecraft began failing. Once again the Soyuz spacecraft could not be stabilized in orbit. Its braking rockets were fired and the capsule entered a stable descent path; however, the heat shield was badly damaged. As if that hadn't been bad enough, during sep-

The Soyuz Spacecraft

Development of the spacecraft with the internal Soviet designation 7K OK (OK is the abbreviation of the Russian word for Orbital Ship) began in the early 1960s, as a counterpart to the American Gemini capsules. It was given the name Soyuz (Union).

Among other things, the Soyuz was used to test flight and docking maneuvers in preparation for the Soviet lunar mission. The first-generation Soyuz spacecraft were used from 1967 to 1971, primarily to transport crews to and from the Salyut space station.

The early Soyuz spacecraft consisted of three units: the orbital module, the descent or landing module, and the service module. It had an overall length of about 23 feet. The orbital module accounted for 8.5 feet of this and weighed 2,865 pounds. The landing module, weighing 6,393 pounds and measuring 6.9 feet long, contained the seats occupied by the cosmonauts during ascent and landing. The two modules measured 7.22 feet in diameter and were supplied with a breathable atmosphere. Weighing 5,732 pounds with a diameter of 8.85 feet, the instrument and landing module contributed 8.2 feet to the spacecraft's overall length.

The crew of up to three cosmonauts had a "living space" of about 353 cubic feet. Except during preparations for and execution of extravehicular activities, throughout the entire flight the crew wore comfortable clothing instead of spacesuits. Unlike the American Gemini spacecraft, whose energy was produced by batteries and fuel cells, the Soyuz spacecraft was supplied with electrical power produced by two solar panels. The folded panels were located in the service module to save space and were deployed once the spacecraft was in orbit. With the panels deployed, the spacecraft had a span of almost 36 feet from the tip of one panel to the other.

The docking of two Soyuz spacecraft or one Soyuz spacecraft with a space station was accomplished by an automatic-approach radar system called Igla. The docking system itself existed in two variants: the active and the passive adapters. There were thus two types of spacecraft, passive and active. The passive spacecraft waited without any contribution on its part for its actively controlled partner to approach and dock. In the beginning the docking adapter offered no passageway between spacecraft. Its purpose was solely to mechanically connect the two vehicles. Movement of the crew from one spacecraft to the other was possible only by means of an extravehicular activity

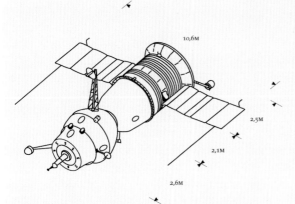

The three modules of a Soyuz spacecraft, type Soyuz 7K-OK.

Soyuz 7K-OK in flight configuration.

and was therefore time consuming and dangerous. Only the last two spacecraft of this series, designated 7K OKS, had an improved docking system offering the ability for crew transfers without extravehicular activities.

With the 7K OK capsules, the Soviet Union achieved several firsts in the history of space travel. In 1967, two Soyuz capsules carried out the first docking maneuver between two unmanned spacecraft, followed two years later by the first manned docking. During the maneuver, two cosmonauts moved through space from one vehicle to the other. In 1970, a Soyuz carried out the longest manned spaceflight to date, lasting eighteen days. This type of spacecraft was also involved in two fatal accidents, however-er. The accidents involving Soyuz 1 and Soyuz 11, in 1967 and 1971, cost the lives of four cosmonauts.

Over the years and decades the Soyuz 7K OK underwent a continuing development process without any fundamental change in its basic structure. The subsequent development stages were the Soyuz T, Soyuz TM, and Soyuz TMA. The Soyuz TMA is still in service today, delivering crews to the International Space Station.

aration of the orbital and landing modules the cabin atmosphere escaped. The capsule crashed into the frozen Aral Sea and sank to the bottom after breaking through the ice. Recovery took several days.

Although the Russian engineers were struggling with numerous design problems, the Soviet leadership decided to send a man into space in the new space capsule. Many involved in the program harbored serious doubts as to whether the spacecraft was ready for a manned flight. But because of the repressive conditions that then existed in the Soviet Union, most dared not openly express these doubts, since critics could quickly find themselves in the gulag. Leonid Brezhnev, then the head of state of the Soviet Union, obviously exerted influence.

With the upcoming May Day celebration, one of the most important socialist holidays, he demanded a space mission be carried out that would receive favorable media coverage.

Despite the technical uncertainties, the project leadership planned an ambitious scenario: a double launch by two Soyuz spacecraft, a docking maneuver between the two, and the transfer of two crew members from one spacecraft to the other as part of an extravehicular maneuver.

The launching of the two spaceships was initially planned for April 23 or 24, 1967. The four-day mission was to begin with the launch of Soyuz 1, the active partner in the docking maneuver. Its pilot would be forty-year-old cosmonaut Vladimir Komarov. At the age of just fifteen, Komarov had begun pilot training with the

Vladimir Komarov.

Vladimir Komarov with his close friend Yuri Gagarin while hunting.

Soyuz 1 on the launchpad.

armed forces on the eve of the Second World War, and in 1960 he became one of the Soviet Union's first group of twenty cosmonauts. Four years later he went to space for the first time as part of the three-man crew of the one-day maiden flight into space by a Voshkod spacecraft.

The flight of Soyuz 1 would be his second space mission. His backup was his close friend Yuri Gagarin, the first man to go into space. The next day, Soyuz 2 was to follow carrying the three cosmonauts: Valeri Bykovsky, Yevgeny Khrunov, and Aleksey Yeliseyev. After the docking, Khrunov and Yeliseyev were to transfer from Soyuz 2 to Komarov's capsule and return to earth with him.

The launch of Soyuz 1 was finally set for 0335 Moscow time on the morning of April 23, 1967. Soyuz 2 was to follow the next day at 0310.

At 0300 Vladimir Komarov boarded his space capsule, clad only in a light-gray flight suit made of wool and a blue jacket. Launch preparations were uneventful and the booster rocket lifted off precisely at the planned time.

After 540 seconds, Soyuz 1 achieved orbit. Komarov was the first Russian cosmonaut to make a second spaceflight.

TASS, the Russian news agency, made a brief report on the successful launch of Soyuz 1 in its familiar mystery-monger style. Only vague information was given as to the purpose of the mission. Only a few words were dedicated to the coming launch of Soyuz 2.

Meanwhile, telemetry received from Soyuz 1 indicated a problem onboard the spacecraft. One of the two solar panels had obviously failed to deploy. Equally critical was that exhaust gases from the engines had fouled an optical star sensor during launch. It was thus questionable whether Soyuz 1 would be able to carry out the flight attitude maneuver required for the docking maneuver. After initial communications problems—the capsule's shortwave radio did not work—VHF communication was finally established with Komarov. In a calm voice he reported that he felt well, and confirmed the telemetry information.

After several unsuccessful attempts to point the one remaining solar panel at the sun, the spacecraft's energy situation was extremely poor. It was sufficient only for a short flight, but not for the extensive flight program that had been planned. Komarov was instructed to turn off all electrical devices that were not absolutely necessary, in order to save power.

Despite the tense situation, the decision makers still had hopes that they could clean up the situation and carry out the docking maneuver. During orbits seven to thirteen, Soyuz 1 was outside VHF radio range. Before contact was lost, Komarov was instructed to use the time to get a few hours of sleep, while in the control room, hectic work went on to find a solution to the technical problems. Consideration was even given to sending Soyuz 2 to help. According to the plan, cosmonauts Khrunov and Yeliseyev would carry out an extravehicular mission to try to manually free the jammed solar panel of Komarov's capsule.

When contact with Komarov was restored after the thirteenth orbit, the situation remained critical. The launch of Soyuz 2 was canceled.

Simultaneously the control center decided to begin the reentry of Soyuz 1 during its seventeenth orbit. Komarov was ordered to end his so-far-unsuccessful attempts to point the solar panel at the sun. His remaining fuel was more urgently needed for ignition of the braking rockets. With the loss of the star sensor and the jammed solar panel, manual control by the cosmonaut seemed to offer the only chance. The beginning of the reentry would have to take place in the darkness of the earth's shadow, which was viewed as particularly problematic. Komarov pointed out that orientation would be very difficult for him. He would have to carry out a series of complicated maneuvers to place the spacecraft in the correct attitude for the braking maneuver before entering the earth's shadow. After leaving the darkness he would have to check to determine if the ship had retained its attitude. If that was not the case, he would have to make another attempt to right the spacecraft. This was an extremely complex process and had never been practiced in this form during training. The matter was further complicated by the fact that the maneuver would have to take place outside radio range of the ground stations. Komarov would have to act without help. Yuri Gagarin insisted that he be permitted to give his friend this information himself.

At precisely the calculated moment, Komarov tried to ignite the braking engine. It failed. The Soyuz control computer had obviously prevented the ignition command because of the lack of attitude information resulting from the defective star sensor. A second attempt was planned for the nineteenth orbit. Meanwhile, Komarov seemed calm and composed, at least outwardly. He carried out his tasks absolutely correctly and precisely at the scheduled time.

The moment of impact with the ground, photographed from a recovery helicopter.

The braking maneuver, which was to last 150 seconds, was scheduled for 0557 hours and 15 seconds on April 24. This time, Komarov succeeded in igniting the engine almost to the precise second. He radioed to earth: "The engine worked for 146 seconds. Shutdown at 0559 hours and 38.5 seconds . . . Alarm 2." Alarm 2 meant that when the recovery engine was fired, the spacecraft had deviated more than 8 degrees from its predetermined attitude and was therefore plunging toward earth out of control on an unplanned reentry path.

Fifteen minutes later the standard and expected break in communications occurred. As it entered the atmosphere, Soyuz 1 was shrouded by ionized gases, which absorbed all radio waves. Moments later, Komarov's voice was again heard from the ground control loudspeakers. Despite the dramatic situation, he sounded calm and unexcited.

Meanwhile, from the flight data the control center determined the probable landing site.

Search and rescue teams were mobilized. In the predicted landing area it was a beautiful sunny morning, and visibility was excellent. Finally the pilot of a search aircraft reported that he had been able to make out the capsule in the air.

The next report came from the crew of a recovery helicopter: it had discovered the capsule lying on its side in a green meadow, and the parachute was spread out beside it. But then the helicopter crew saw something very unusual. The thrusters, which should have fired fractions of a second before landing to drastically reduce the capsule's impact speed, had just ignited, after the capsule was already on the ground. This was of course too late to perform their intended function.

The pilot of a search aircraft radioed a disturbing message: "I see the object; the cosmonaut urgently requires medical assistance." The communications between the recovery team and the control center were then suddenly interrupted, probably to avoid giving any clues to interested listeners. It would be three

A recovery helicopter has landed near the accident site.

Burning wreckage.

The Soyuz landing module has collapsed in upon itself.

and a half hours before information about the fate of the cosmonaut reached the control center.

The first recovery helicopter landed just 300 feet from the landing site. The capsule was enveloped in thick black smoke, with the heat shield completely burned away. A fierce fire raged in the interior of the Soyuz. Witnesses later said that they had observed streams of molten metal. The men of the recovery team fought the fire with fire extinguishers and even shoveled dirt onto the blaze. While the desperate effort to extinguish the fire was still going on, the capsule completely collapsed. All that was left was s pile of wreckage.

Vladimir Komarov had to be dead, although initially his body could not be seen in the remains of the spacecraft. Finally, the completely charred body of the cosmonaut was discovered and removed from the wreckage of the spacecraft.

Residents of the area near the landing site reported that the spacecraft had plunged toward the earth at high speed. The parachute had not completely opened and had spun continuously.

State funeral for Vladimir Komarov in Red Square in Moscow.

They confirmed that the braking thrusters had not fired. After impact, explosions were seen and heard, and then the capsule had begun to burn. Soyuz 1's hapless flight had lasted one day, two hours, forty-seven minutes, and fifty-two seconds.

Toward noon the head of the party and state, Leonid Brezhnev, who was at a Communist Party conference in Czechoslovakia, was informed about the disastrous end of the mission. After more then twelve hours of silence, the TASS news agency announced that Komarov had been killed on landing. Officially, it was said that the flight had initially gone according to plan. While returning to earth the parachute had opened at an altitude of 23,000 feet. According to the report, the spacecraft subsequently struck the ground at high speed. It was purported that the landing parachute had failed to function properly, causing the accident.

The remains of the cosmonaut were delivered to Moscow and were cremated there. The urn containing his ashes was displayed in the central building of the Soviet army. Many dignitaries and thousands of Soviet citizens paid their last respects to Komarov. On April 26, Komarov's physical remains were interred in the Kremlin Wall Necropolis in Moscow.

As an initial reaction to the accident, the Soviets canceled all manned spaceflights until further notice. On June 20 of that year, an investigation committee created by senior state officials delivered its report. The cause of the crash was given as failure of the main parachute release mechanism.

During the flight the parachute was stowed in a container. During reentry into the denser atmosphere, a drogue chute was first deployed. This slowed the capsule until its rate of descent fell below 450 miles per hour. Then the auxiliary parachute had to pull the main parachute from its container. It was then supposed to deploy, fill with air, and further slow the capsule's descent. Finally, just a few feet above the ground, solid-fuel rockets ignited and reduced impact speed to a level that those inside could tolerate. So much for the theory.

Komarov's widow during the interment of his urn in the Kremlin Wall.

In the case of Soyuz 1, the auxiliary parachute failed to extract the main parachute from its container. Somehow or other it had become stuck. The space capsule's control system reacted to the still-increasing descent rate and gave the command to jettison the obviously malfunctioning parachute. But since it had initially remained in its housing for unexplained reasons, this had no effect. The spacecraft remained connected to the braking parachute, still in its container; the braking parachute was located above the capsule. The reserve emergency parachute was released by the automatic control system and began to deploy, but its lines obviously became tangled with those of the still-open drogue parachute. The reserve parachute did not fully deploy, and its braking effect did not reach what was required for a safe landing. As a result the spacecraft struck the ground at the much-too-high velocity of more than 87 miles per hour. An autopsy of Komarov's physical remains confirmed this scenario: the cause of death was injuries from the severe impact; the serious burns were not caused until after the cosmonaut's death.

The failure of the parachute was caused by sloppiness on the part of the technicians who had prepared Soyuz 1 for the mission. Before launch, the spacecraft had been fitted with a thermal protective cover. During this procedure the parachute container was empty, with its cover open. As a result, the overcoat had also found its way into the container, and there formed a rough surface. The friction between this outer surface area and the parachute that was later squeezed into the container had been so great that the braking parachute's pulling force had been inadequate to extract the main parachute from its container.

Unfortunately, this meant that the sister ship, which had been prepared in the same way, and its three-man crew would very probably have met a similar fate on landing as had Komarov. The three men had been saved only because Soyuz 1's technical problems had prevented the launch of Soyuz 2.

To this day, there are various rumors about the circumstances of Vladimir Komarov's death. It is claimed that, aware of his unavoidable

death, he spoke by radio to government representatives and his wife before beginning reentry. Sources report that he criticized, even cursed, those responsible during his plunge to earth. Supposed proof is available on the Internet, but its authenticity is more than questionable, however. Not until after the opening of the Soviet Union to the West did reliable facts about the events of April 1967 gradually come to light of day, a summary of which forms this account.

Komarov's name and those of thirteen other astronauts and cosmonauts who lost their lives are on this plaque, which the Apollo 15 astronauts left in the lunar dust after their landing on the moon in 1971.

NERVES LIKE WIRE ROPE: GEMINI 6A

On October 12, 1964, the Soviet Union launched the 5.5-ton spacecraft Voshkod 1, carrying a crew of three. At that time, NASA had just completed six missions with the rather minimalist Mercury one-man capsule. The program had ended in May 1963, with the thirty-four-hour flight of Mercury-Atlas 9 carrying astronaut Gordon Cooper.

Since then the United States had been working on a follow-up program with the two-seat spacecraft named Gemini (Latin for twins). During preparations for the first Gemini flight, the sensational news broke of the first human to leave a spacecraft in space. Carrying a crew of two, Voshkod 2 had been launched on March 18,

1965. Soon after it achieved orbit, cosmonaut Alexey Leonov climbed out for a spacewalk and, connected to the spacecraft by a thin umbilical cord, spent about twenty minutes in space. This was another bitter setback for the Americans.

The purpose of the first Gemini flights was to extend the time that the astronauts spent in space, and to investigate the effects of extended weightlessness on the human body. The first manned mission in the program was also the first manned American flight after a break of almost two years. Gemini 3 was launched on March 23, 1965, the flight lasting just five hours.

Gemini 4, the second mission, included the first extravehicular maneuver by an American, the astronaut Ed White. And the next flight by Gemini 5 was of considerably longer duration, almost eight days.

Pilot Tom Stafford (left) and Commander Walter "Wally" Schirra. *NASA*

NASA-S-65-893

The Gemini spacecraft.

Gemini

In December 1961, NASA unveiled a follow-up program to the Mercury missions, which was to test the methods and procedures for a landing on the moon. The two-seat spacecraft with which the United States intended to achieve this was called Gemini, after the constellation of the same name.

The Gemini spacecraft was developed from the single-seat Mercury capsule. With an overall length of 19 feet, the spacecraft was made up of two parts: a return capsule weighing 4,850 pounds and an adaptor ring weighing 2,425 pounds. The latter connected the capsule to the booster rocket during launch and provided the space vehicle with energy throughout the entire mission. It also housed thrusters for changing the capsule's flight path and their fuel tanks. The adapter

ring was no longer needed after the braking maneuver for reentry into the earth's atmosphere and was jettisoned.

The crew cabin was bell shaped with a towerlike attachment in which the landing parachutes, signaling equipment, and radar equipment necessary for rendezvous missions were housed. In the crew cabin the two astronauts sat side by side. Each had his own entry and exit hatch with a tiny window.

The Gemini space capsules were launched by Titan booster rockets. Unlike the Mercury spacecraft, they were not equipped with an escape tower. In the event of an accident during launch, the astronauts were supposed to be catapulted out of the capsule by means of ejection seats. This was not entirely without risk, however. An ejection un-

der such conditions would expose the astronauts to a force of up to 20 G, twenty times
their body weight.

During the period from March 1965 to
November 1966, sixteen astronauts flew a
total of ten missions in space, four men even
going twice. Five astronauts left the safety
of their spacecraft during nine extravehicular
activities, or spacewalks. They spent up to
five and a half hours outside their capsules,
working in open space.

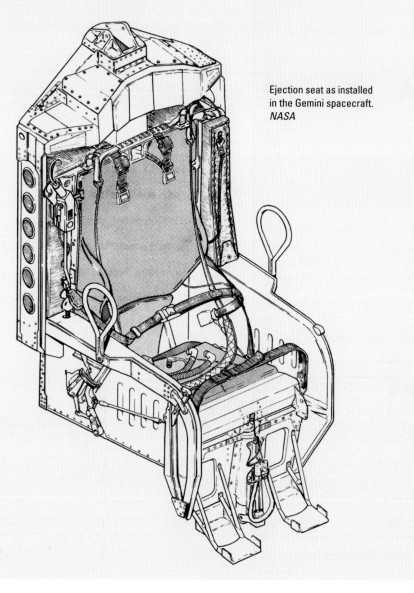

Ejection seat as installed
in the Gemini spacecraft.
NASA

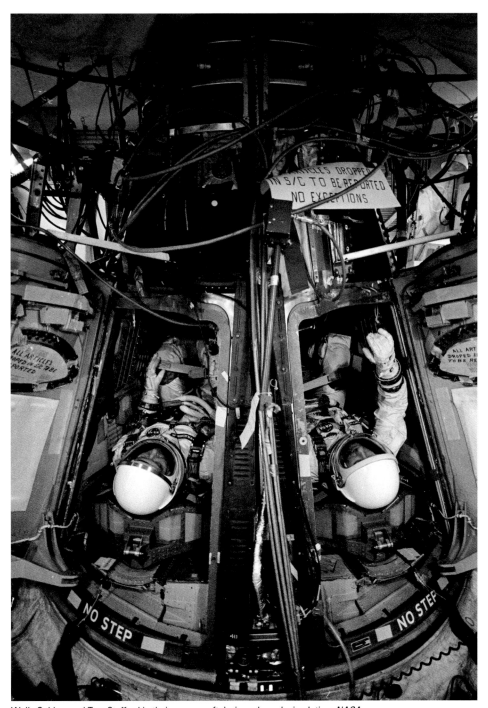

Wally Schirra and Tom Stafford in their spacecraft during a launch simulation. *NASA*

The engines of the Titan booster rocket have ignited, only to be switched off seconds later. A large cloud of smoke spreads over the launch site. *NASA*

Because the main objectives of the Gemini program were to develop and test the techniques and procedures for the flight to the moon, the next flight, Gemini 6, was to see the first approach to an orbiting unmanned target satellite. The high point of the mission was to be a docking with this Gemini-Agena Target Vehicle (GATV). The grew of Gemini 6 under the command of Mercury veteran Walter "Wally" Schirra and space rookie Tom Stafford painstakingly prepared for the launch on October 25, 1965.

Wally Schirra, a married, forty-two-year-old naval aviator with two children, was one of the first group of American astronauts, the Mercury Seven. In 1962, he flew in a Mercury capsule, orbiting the earth six times in nine hours. The flight of Gemini 6B would be his second space mission. Air Force pilot Tom Stafford's flight in Gemini 6A would be his first trip into space. Like Schirra, he was married and the father of two children. Since 1962, he had been a member if the second group of NASA astronauts.

The target satellite GATV 6 was supposed to be launched into space just a few minutes prior to their own launch—however, it failed. The booster rocket exploded six minutes after liftoff, and the GATV was destroyed. Schirra and Stafford, already waiting in their capsule to follow the GATV into space, had to leave their spacecraft empty-handed.

Under pressure from the successful Soviet spaceflights, the plan was now developed to attempt a formation flight by two Gemini capsules instead of a docking maneuver with an unmanned satellite. To define the flight of the Gemini 6 mission, because of the changed mission profile, it was renamed Gemini 6A. The target for Gemini 6A was to be Gemini 7, carrying astronauts Frank Borman and Jim Lovell. After this mission, NASA inexplicably switched to Roman numerals for the numbering of flights—for consistency and better readability, however, we will continue to use Arabic numbers.

Gemini 7 in fact launched on December 4, 1965, on a long-term mission in space lasting almost fourteen days. Gemini 6A was to follow eight days later on December 12. The countdown went flawlessly on that day. The Titan booster rocket's engines ignited precisely at the predetermined time of 0945 local time, but a few seconds later they shut down. The Titan sat motionless on the launchpad. Those watching live on television held their breath.

An extremely critical situation had developed before the eyes of the viewers. Had the rocket lifted off from the launchpad by even a few inches, after the engines shut down it would have fallen back like a stone. It would probably have collapsed, and with its more than 140 tons of highly explosive rocket fuel it would have been destroyed in a terrific explosion. In this situation, the regulations called for Commander Schirra and his copilot Stafford to use their ejection seats to escape the area of the expected fireball.

Schirra had watched closely as the clock in the cockpit began to run, showing that the flight had begun. All the same, however, he had felt no movement of the rocket and the space capsule, causing him to deduce that the Titan had been unable to leave the launchpad. He quickly made the highly risky but in retrospect the only correct decision. He decided not to use his ejection seat, and to remain in the capsule to wait for what was to happen. Tom Stafford, who could also have used his ejection seat, kept his nerve just as well as Schirra.

It turned out that a cable had separated due to vibration caused by the rocket engine running up. The purpose of this cable was to close

Ejection seat in a Gemini capsule. *NASA*

a contact on liftoff, to start the cockpit clock and indicate the flight duration to the crew. Since the clock was already running but the rocket had not yet moved, the launch computer identified it as a failure and abruptly shut down the engines.

The ice-cold reaction by the two astronauts had saved NASA from a costly and, for the astronauts, a potentially risky exit from the space capsule. Not only would the inner workings of the space capsule have been irreparably destroyed, during ejection the astronauts would have been exposed to forces twenty times greater than their body weight and would very likely have suffered serious injuries. One way or another, Schirra and Stafford had to leave their spacecraft empty-handed for a second time.

A subsequent examination of the Titan booster rocket discovered additional problems, and not just more poorly plugged-together ca-

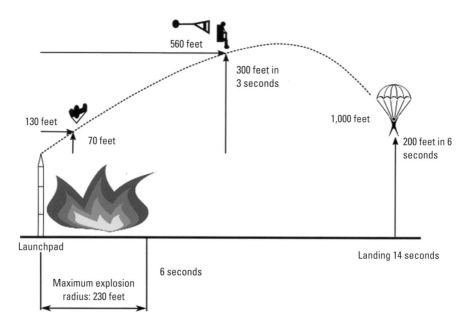

560 feet

300 feet in
3 seconds

130 feet

1,000 feet

70 feet

200 feet in 6
seconds

Launchpad

Landing 14 seconds

6 seconds

Maximum explosion
radius: 230 feet

Escape by ejection seat was a traumatic and quick action. *Woydt*

ble connections. Inside one of the engines, a dust cover was found that had been forgotten during assembly of the rocket and had obviously been overlooked during launch preparations. No one really wanted to envision what effect this might have had.

The third launch attempt, three days later on December 15, was finally successful. Gemini 6A and Gemini 7, which had been in orbit for eleven days, approached to within a foot of one another. The spacecraft matched speeds, flew alongside one another, and performed fly-arounds. The mission objective had been achieved. Schirra and Stafford had accomplished the first human-controlled rendezvous maneuver in earth orbit.

Gemini 6A returned to earth after just one day in space. Gemini 7 followed after thirteen days and eighteen hours in earth orbit and, on December 18, 1965, splashed down in the Atlantic. Borman and Lovell had thus broken the existing endurance record in space. Their record would hold for five years, when it was surpassed by the flight of the Soviet Soyuz 9, which lasted more than seventeen days.

EMERGENCY LANDING IN THE PACIFIC: GEMINI 8

The Gemini Program had already achieved two of the major goals of the American space program: an astronaut, Ed White, had left his spacecraft and spent several minutes floating in space alongside it. This had served as proof that a man could in fact exist in the hostile environment of space and could orient himself in the unfamiliar environment of weightlessness. And finally, the crew of Gemini 7 had spent almost fourteen days in space, roughly equal to the duration of a lunar flight.

Far up the list of things yet unaccomplished was the docking of two spacecraft. Gemini 6A and Gemini 7 had approached to within a foot of one another and carried out joint flight maneuvers. But the last step was still missing to make the matter perfect. Gemini 8 was now to repeat what Gemini 6 had been prevented from doing by the failed launch of the Gemini-Agena Target Vehicle (GATV): the docking of a Gemini spacecraft with an unmanned target satellite. During its mission, Gemini 8 was to carry out four docking and separation maneuvers in succession with one and the same GATV.

Commanding the mission was Neil Alden Armstrong, who was the first civilian to go into space. The thirty-five-year-old Armstrong was married with two children. During his time as a test pilot at Edwards Air Force Base, his activities included flight testing the famous X-15 experimental aircraft, and in 1962 he had been chosen to join the second group of American astronauts. Gemini 8 would be his first space mission.

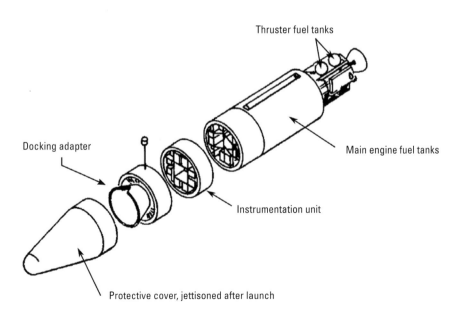

Thruster fuel tanks

Docking adapter

Main engine fuel tanks

Instrumentation unit

Protective cover, jettisoned after launch

Schematic drawing of a Gemini-Agena target satellite. *NASA/Woydt*

The crew of Gemini 8: Dave Scott (left) and Neil Armstrong. *NASA*

Air Force pilot David "Dave" Scott was selected to be Armstrong's copilot. A major in the air force, the thirty-four-year-old was also married with two children. He joined the third group of American astronauts in 1963.

In addition to the docking maneuver, the flight plan for the three-day mission included an estimated two-hour spacewalk by copilot Scott. During the ambitious operation, Scott was to climb out of the spacecraft after the first of four docking maneuvers. He was to move to the nose of the spacecraft and recover the material experiment package stowed there. Then he was to move over to the Agena target satellite and activate a micrometeorite experiment .

The tightly scheduled plan also envisaged checking several special tools for their effectiveness in weightless conditions. While the astronaut was carrying out his tasks, Commander Armstrong would again separate the Gemini capsule and the target satellite. Scott's next task would be to put on a new backpack (Extravehicular Support Package or ESP) for use during extravehicular activities. Not only was the ESP an independent life support system, it also included a sort of thruster gun. By firing short bursts of

Freon gas, Scott would be able to move in space. To don the ESP, he had first had to reach the rear of the Gemini spacecraft, where the equipment was stowed during launch. While Armstrong increased the distance between the Gemini capsule and the Agena target satellite to up to 60 feet, Scott was supposed to try to move back and forth between the vehicles. But this was not to take place.

Armstrong and Scott clambered into their seats inside the spacecraft at 0945 Florida time on March 16, 1965. Fifteen minutes later the GATV with construction number 5003 was launched from a neighboring launchpad at Cape Kennedy. This time all went well. The GATV achieved a near-circular orbit at 185 miles above the earth.

Precisely forty-one minutes later, Gemini 8 followed the GATV into space with a picture-book launch. The lowest point in its eccentric orbit was 99.4 miles above the surface of the earth; the highest, 169 miles. Armstrong and Scott completed five course correction maneuvers to match the flight path of the GATV. Then they began their approach.

The rendezvous radar located the target satellite from a distance of more than 205 miles. When Gemini 8 had closed to within 87 miles, the crew was able to see it as a bright spot in the blackness of space. During the last 60 miles of the approach flight, the astronauts let the onboard computer control the spacecraft.

When the distance to their target had shrunk to less than 150 feet, Commander Armstrong assumed manual control. First he cautiously maneuvered the capsule around the GATV. Looking through their tiny windows, the astronauts made a visual inspection of the satellite. Everything seemed to be in perfect order.

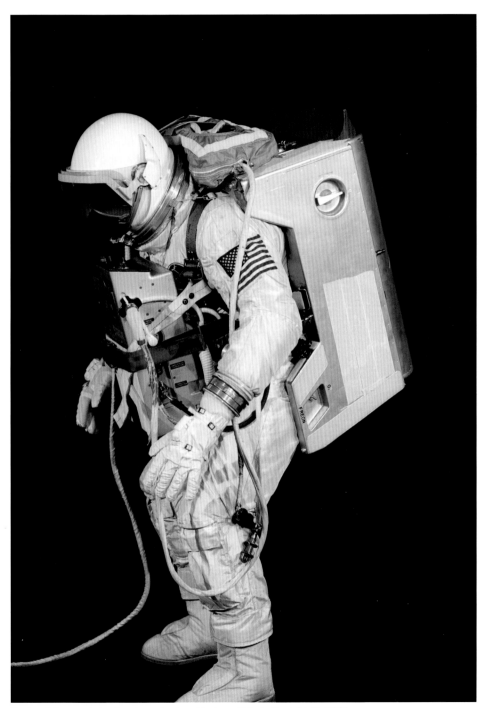

Backpack for use during the planned extravehicular activity by Dave Scott. *NASA*

Launch of the Gemini-Agena satellite, which was to serve as a docking target for Gemini 8. *NASA*

The GATV's docking adapter was on its front end. Seven hydraulic shock absorbers would absorb the kinetic energy produced when the two spacecraft came together. The Gemini's cone-shaped docking adapter had to be guided into its counterpart on the GATV. Once this was done, clamps locked the two adapters together. Electric motors then pulled the cone back, establishing a mechanically stable connection between the Gemini capsule and the GATV. The electrical connection en-

Launch of Gemini 8 on March 16, 1966. *NASA*

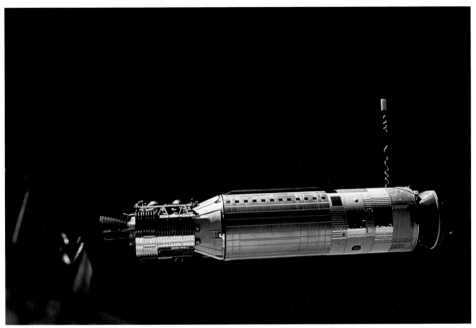

Close-up of the Gemini-Agena target satellite, taken from a window of the Gemini 8 spacecraft. *NASA*

abled the pilot to control the target satellite's onboard system remotely from his position in the Gemini capsule.

After having approached their target for six and a half hours, the crew of Gemini 8 was given the OK to dock. Armstrong approached the docking adapter with short bursts from the thrusters. Finally, contact was established. The two vehicles had come together with a residual speed of just 5.6 miles per hour. The locking mechanism engaged and the first docking maneuver between two spacecraft in space had been achieved. Armstrong reported: "Flight, we are docked." Finally, NASA had achieved a first and had beaten the Soviets.

Houston responded with an instruction with prophetic contents: "If you run into trouble and the attitude control system in the Agena goes wild, just . . . turn it off and take control

with the spacecraft." Then Gemini 8 went out of range of the ground station.

In orbit, Scott issued a command for the GATV to turn 90 degrees, or, in aviation parlance, to yaw 90 degrees. While the satellite executed the command, the astronaut closely watched the Gemini capsule's instruments. He was astonished to see that the vehicles had obviously begun a rolling motion about their longitudinal axis. Since they had just flown over the nightside of the earth, outside the windows it was pitch black. It was therefore impossible to determine if this corresponded to the facts or if the instruments were giving the two astronauts inaccurate information.

Armstrong tried to counter the roll with the Gemini capsule's maneuvering thrusters (OAMS), and at first this seemed to work. But just minutes later the rolling movement began

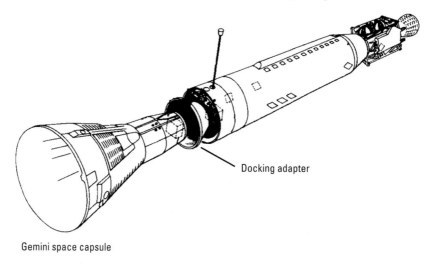

Gemini-Agena target satellite

Docking adapter

Gemini space capsule

Schematic drawing of the docking maneuver. *NASA*

again. After brief consideration, the two astronauts came to the conclusion that the unexpected roll must be due to a malfunction of the GATV's attitude control system. Scott switched off the target satellite's control system as ground control had instructed, which seemed to solve the problem. For the next few minutes the two spacecraft flew without incident.

Finally, Armstrong tried to align the spacecraft's attitude with the horizon by using the maneuvering thrusters. The undesired roll movement immediately began again, this time with a much-higher roll rate than before. The astronauts were puzzled until Armstrong noticed on the displays on his instrument panel that the fuel tank contents gauge showed just 30 percent thruster fuel remaining. From this reading and the rolling movement, he concluded that one or even several of the thrusters must be operating in continuous mode, causing the unusually high consumption of fuel. How-

ever, which of the sixteen thrusters was it? There was no time for a lengthy search. Commander Armstrong decided on a radical solution—he ordered Scott to separate the two space vehicles. Scott acted immediately and Armstrong backed the Gemini capsule away from the GATV. The maneuver did not have the desired effect, to say the least. To their astonishment the roll rate became even higher.

Gemini 8 was now again within radio range of ground control. Telemetry data had shown that the two spacecraft had separated. Scott reported: "We have serious problems here . . . we're tumbling end over end up here." And Armstrong added: "We're rolling up and we can't turn anything off. Continuously increasing in a left roll."

The rate of rotation became ever higher, and finally the spacecraft was whirling at 360 degrees per second. This equaled or exceeded the crew's physiological limits, and Armstrong

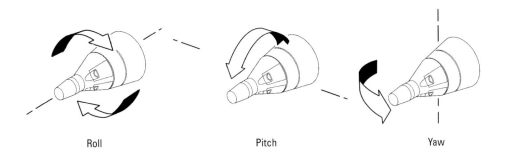

| Roll | Pitch | Yaw |

Space axes of a spacecraft. *Woydt*

and Scott became dizzy. Their field of vision narrowed markedly, a physiological effect that indicated an impending loss of consciousness. Unconsciousness would undoubtedly have had fatal consequences for the astronauts. With the last of his strength and concentration, Armstrong finally shut off the space capsule's attitude control system and activated the independent emergency system, which had originally been intended for use during reentry. The commander's decision appeared to have been absolutely correct: the Gemini capsule reacted to Armstrong's efforts to counter the rolling movement. The rolling stopped and the spacecraft's attitude stabilized.

After he had succeeded in regaining control of the situation, the two astronauts began searching for the cause. One after another, they switched each of the sixteen thrusters on and off again. Using this method, they discovered that thruster #8 had failed to shut off. The cause of the problem had been found. The engine had been in continuous operation and had caused the high rate of roll.

Use of the attitude control backup system caused another problem, however. The mission rules prescribed that in such a case the mission should be aborted immediately. There was

therefore nothing to discuss, and because of the uncertainty of the situation, no one wanted to question this regulation. Consequently, preparations for landing were begun just ten hours after the launch. Because of flight path mechanics, in the current phase of the flight the original landing area in the Atlantic was unreachable. It would have required another twenty-four hours in orbit for the space capsule to be able to come down in the Atlantic. The alternate landing site was in the Pacific. Before the astronauts began reentry and thus the return to earth, as a precaution ground control instructed each of them to take a seasickness tablet, since heavy seas were expected in the Pacific. Meanwhile, the destroyer USS Leonard F. Mason was dispatched to the landing site, located 500 miles east of the Japanese island of Okinawa.

During their seventh earth orbit, while flying over the central African Republic of Congo on the nightside of the earth, Armstrong and Scott fired the retrorockets. Gemini 8 was now on its way back to earth.

Barely thirty minutes later, a C-54 search aircraft spotted the spacecraft beneath its parachute precisely at the anticipated time and place. Splashdown occurred ten hours, forty-one minutes, and twenty-six seconds after liftoff from Florida. The C-54 dropped three divers by parachute, and the frogmen placed an air-filled stabilization ring around the space capsule, which was dancing in the waves. This gave the capsule additional buoyancy and kept it on the surface until the recovery ships rushing to the scene arrived.

Finally, after three long hours, the Leonard F. Mason came alongside and took the exhausted but uninjured astronauts aboard. The ship took them to Okinawa, from where they set out on the journey home to Cape Kennedy. On their arrival, Neil Armstrong answered a question from a reporter about their experience: "We had a great flight—at least for the first seven hours."

The space capsule was immediately taken to McDonnell Douglas, the company that man-

ufactured it, in St. Louis, Missouri, where it underwent a detailed examination. A short in the electrical system was determined as the most probable cause for the failure of the maneuvering thruster. Pilot error was eliminated as a possible cause. Indeed, the cool and courageous actions of Armstrong and Scott had saved them from a potentially life-threatening situation and enabled them to get their spacecraft safely back to earth. The astronauts had done everything right.

It is very likely that Armstrong's calm and considered action in this critical situation had a decisive influence on his subsequent career as an astronaut. His composure may have been a deciding factor in his selection for the job of his life three years later.

Armstrong and Scott wait for the arrival of the recovery vessel. The space capsule has been secured by divers dropped from an aircraft. *NASA*

DEATH AT SUPERSONIC SPEED: X-15

After the end of the Second World War, it wasn't long before the struggle for world dominance between the United States and the Soviet Union began a period known as the Cold War. The two major powers competed in every imaginable field, and of course in aviation and space travel. The development of newer and ever more advanced weapons systems had the highest priority and was pursued by both sides, which committed vast amounts of people, material, and money.

As early as the 1940s, the United States had created a research program in which new experimental aircraft, so-called X aircraft (with "X" an abbreviation for experimental), would push the limits of and improve aviation development. As part of this program, in 1947, test pilot Chuck Yeager officially broke the sound barrier, until then a seemingly impenetrable physical barrier, for the first time while flying the Bell X-1.

By the beginning of the 1950s, scientists were working on an experimental aircraft capable of much-higher speeds. It was to have a maximum speed in excess of 3,700 miles per hour, greater than five times the speed of sound, which was designated hypersonic. This design work soon attracted the attention of the military. At a time when rocket technology was in its infancy, an aircraft that not only could fly at very high speed but could reach altitudes in excess of 30 miles seemed an interesting alternative to unreliable rockets for delivering bombs to targets deep in enemy territory.

Over time these early theoretical ideas became the X-15 program, a joint undertaking by the US Air Force and NASA.

The X-15 project took the designers into completely new technical territory, and during the design phase they were forced to consider things that were rather secondary in the design of "normal" aircraft. Air friction at high speeds proved particularly problematic. At high speeds the aircraft's outer skin could be heated to temperatures of 1,800 degrees Fahrenheit and more. Control of such an aircraft could not rely solely

An X-15 in flight. *NASA*

The X-15 Hypersonic Aircraft

The X-15 was an experimental aircraft from the X series, used jointly by NASA and the US Air Force. It was used to research flight behavior as extreme speeds at high altitudes. The flights took place in the hypersonic speed range, which meant speeds of more than 3,700 miles per hour, and to the edge of space at heights greater than 60 miles.

The objective of the X-15 program was to investigate control options inside and outside the atmosphere, in particular with respect to spacecraft returning from space. Materials research experiments tried to answer the question of how the immense friction and the resulting tremendous heat that occurred in this speed range were to be controlled. With this goal in mind, it is not surprising that the results of the research program ultimately flowed into the development and construction of the space shuttle years later.

The X-15 was made by the well-known aircraft manufacturer North American Aviation. A total of three aircraft of this type were delivered from its factory in Inglewood, California.

The X-15, with its slender cylindrical fuselage, was 50.6 feet long, had a wingspan of 22 feet, and measured 13.5 feet from the ground to the tip of its vertical tail.

The area of the two stub wings was just 215 square feet. With empty tanks the aircraft weighed 14,991 pounds, while with full tanks, launch weight was up to 33,000 pounds.

As fuel, the aircraft carried (internally) a maximum of 1,004 gallons of deeply cooled liquid oxygen and 1,400 gallons of dehydrated ammonia. In the course of the program, fuel capacity was increased through the addition

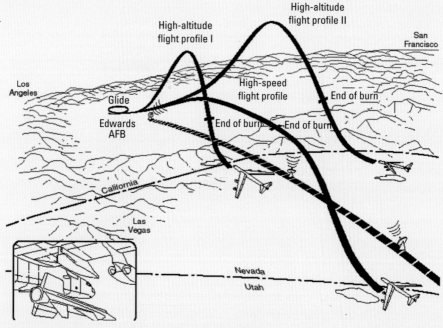

High-altitude and speed profiles from the X-15 test program. *NASA*

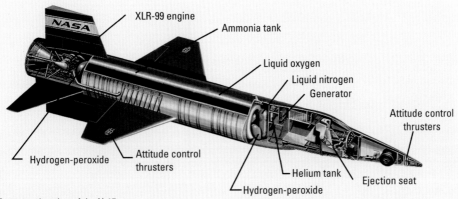

XLR-99 engine

Ammonia tank

NASA

Liquid oxygen

Liquid nitrogen

Generator

Attitude control
thrusters

Hydrogen-peroxide

Attitude control
thrusters

Helium tank

Ejection seat

Hydrogen-peroxide

Cutaway drawing of the X-15

of external fuel tanks. The internal tanks were behind the cockpit, separated from it by an equipment bay for scientific instruments. It was shaped like a hollow cylinder and also formed the aircraft's outer hull. In its central core there was another smaller cylinder. It contained helium under high pressure, which kept the two large main tanks pressurized. The hydrogen-peroxide drove the turbopumps, which delivered oxygen and ammonia to the engine's combustion chamber. Two redundantly designed power generators supplied the aircraft with electric power.

A conventional nosewheel and two steel skids on the aft fuselage formed the retractable undercarriage. Metal skids were used for the main undercarriage instead of wheels, saving space and weight and being more durable than rubber tires. On landing, the X-15's skids touched the ground first. The enormous friction of the skids on the ground caused dramatic deceleration, which caused the nosewheel to slam onto the ground. The pilot had little ability to steer the X-15 after it touched down, and it therefore required a large landing zone. For this reason it did not land at developed airfields with paved run-

ways. Instead the aircraft landed on dry lakebeds, which offered sufficient room.

The X-15 was dropped from a mother ship, a converted Boeing B-52 strategic bomber designated the NB-52, which carried the rocket plane into the air. At an altitude of more than 42,500 feet, the rocket aircraft was dropped by the mother ship at a speed of 500 miles per hour. Moments later, the X-15's rocket engine was ignited and the flight, which normally lasted about ten to eleven minutes, began with one of two possible flight profiles. One was a high-altitude flight, which saw the aircraft make a steep climb and a rather shallower descent. The alternative was a speed flight, in which the X-15 flew horizontally at a predetermined speed.

Within the earth's atmosphere, the X-15 was controlled conventionally like a normal aircraft, with horizontal and vertical tail surfaces. The two unusually thick, wedge-shaped rudders were more than control surfaces. V-shaped opening flaps also influenced the aircraft's flight characteristics. The elevators on both sides of the tail were capable of moving independently of one another. This made it possible to control ascent and

The cockpit of an X-15. *NASA*

descent, but also to control rolling motion by the aircraft. At high altitudes, controllability of the aircraft by aerodynamic control surfaces at first became difficult and then finally quite impossible because of the increasingly thin atmosphere. At that time a change was made to a ballistic control system, and the X-15 was controlled by attitude-control thrusters like a spacecraft.

During initial trials the X-15 was controlled by relatively low-powered XLR11 rocket engines. These were later replaced by the specially designed XLR99, which produced 60,000 pounds of thrust, an impressive figure for such a small and light aircraft.

It operated for a maximum of 239 seconds, after which the fuel was expended.

A total of 199 flights were carried out during the X-15 program, which lasted from 1959 until 1968. Twelve pilots flew the aircraft. There were several accidents, including one with fatal results.

An X-15 just seconds before it is released by the NB-52 mother ship. *NASA*

on vertical and horizontal tail surfaces as on a conventional aircraft. The aircraft had to remain controllable in the thin upper atmosphere, and in extreme cases in the vacuum of space.

Based on extensive studies, a surprisingly simple catalogue of requirements was created for the X-15. Its maximum achievable speed was to be at least 5,300 miles per hour, and maximum service ceiling was 393,700 feet. The aircraft had to be designed for two different flight profiles. One was for high-speed flights at relatively low altitudes, but the aircraft's characteristics also had to permit parabolic flight paths to the edge of space and beyond.

The problem of friction was solved by a new, innovative material for the aircraft's outer skin. Called Inconel, the alloy consisted largely of nickel and chrome and retained its strength over a very large temperature range. The special design of the wings and control surfaces allow the individual components to stretch and shift. The aerodynamic characteristics of the

X-15 thus remained unchanged over a wide temperature range.

The problem of a sufficiently powerful propulsion system could be solved only by a rocket engine. Relatively reliable rocket engines existed for use in intercontinental ballistic missiles, but something new was required for use in aircraft.

Finally, three X-15 aircraft were built under the direction of North American Aviation in its California factory. Reaction Motors Inc., which had been involved in the project since 1959, was responsible for development of the propulsion system. The project was based at Edwards Air Force Base in California, and all of the total of 199 flights made by various X-15 aircraft were carried out from there.

Regardless of its mission, every X-15 flight followed the same routine. The mother ship, a specially modified B-52 bomber, carried the X-15 to the launch altitude of 42,650 feet, at which height the aircraft was released.

Ground personnel with an X-15 after it landed on a salt lake. *NASA*

After a few seconds the pilot ignited the X-15's rocket engine and climbed to the planned altitude at the desired speed. The powered phase of the flight ended when the aircraft's fuel was consumed. The pilot carried out scientific experiments and executed predetermined flight maneuvers. If everything went according to plan, the aircraft subsequently landed on the hard, level surface of a dry, salt lakebed on the vast grounds of Edwards Air Force Base.

In July 1963, an X-15 set a new altitude re-

cord for manned aircraft at 354,199 feet. The fastest flight of the entire program took place in October 1967. The maximum speed achieved by the aircraft was an amazing 4,520 miles per hour. It was a record for a manned aircraft in the atmosphere, holding for many years, and was broken only by a space shuttle returning from space. The US Air Force set the boundary of space at 50 miles, and on the basis of this American definition, eight of the twelve X-15 pilots became astronauts by reaching or surpassing this mark. The international definition, however, defines an altitude of 62 miles as the boundary of space. Using this measurement, just one of the twelve X-15 pilots can be considered an astronaut.

Not all the flights carried out until mid-1967 went according to plan. In addition to various technical problems, which the pilots were able to master, there were also several crash landings. The aircraft suffered varying degrees of damage, but in each case were repaired and made operational again. All the pilots survived with more or less serious injuries.

This changed on November 15, 1967. The pilot on that day was Michael "Mike" Adams, a thirty-seven-year-old Air Force test pilot. He had made his first flight in the X-15 in 1966, and by this day had flown the type a total of six times on missions with different flight profiles.

This was the 191st flight of the X-15 test program. After the aircraft was dropped by the mother ship, it was to fire its engine for seventy-nine seconds and accelerate to more than five times the speed of sound while climbing to 249,000 feet. The plan called for scientific experiments such as measuring the sun's spectrum and collecting micrometeorites, to be carried out during the glide back to Edwards Air Force Base. A material-scientific experiment was to investigate the behavior of materials to be used in construction of the Saturn V lunar rocket.

With the X-15 slung beneath its starboard wing, the mother ship lifted off the runway at Edwards Air Force Base at 0912 California time and headed north. The NB-52 climbed to the release altitude of 45,000 feet. After a short delay—an observer aircraft had not yet reached its designated position—at 1030 + 7 seconds the X-15 was released. Mike Adams started the rocket engine within one second of

Michael Adams in a pressure suit beside his X-15. *NASA*

being dropped, and just 1.5 seconds later the rocket engine was operating at full thrust.

During the rocket-propelled flight phase, a problem developed in the X-15's electrical system. The control system was briefly affected; however, with the aid of the backup system, Adams was able to get around the problem and continue the flight as planned.

The engine operated for 82.3 seconds, which was 3.3 seconds longer than planned. This caused the ceiling of the parabolic flight path to be 265,750 instead of 249,300 feet, which was 16,400 feet higher than originally planned. Ultimate speed was 3,617 miles per hour.

Just as the aircraft flew through the apex of the parabolic flight path, the aircraft's nose began turning left from the direction of flight. The aircraft began to yaw. Maximum deviation was 15 degrees, and then the aircraft temporarily returned to its original attitude, only to yaw to the left again shortly afterward. Within sixty seconds of passing the 229,660-foot mark, the aircraft turned 90 degrees to the left—the X-15 was thus oriented at right angles to the direction

Michael Adams is strapped into the cockpit of an X-15. *NASA*

of flight. The only reason that the aircraft was able to survive this extreme flight attitude was the thin air at that altitude. Even at five times the speed of sound, friction had no significant impact on the X-15's flight characteristics.

Then, however, the situation got out of hand for Mike Adams. The pilot advised flight control: "I'm in a spin." He was talking to William "Pete" Knight, another X-15 pilot, who was handling communications that day, but he didn't seem to have understood correctly. He instructed Adams to pay attention to his angle of attack: "Watch your angle of attack, Mike." Instead of an answer, Adams repeated, "I'm in a spin!"

Adams fought desperately to control his tumbling aircraft. In addition to the aerodynamic controls—elevators and rudder—he simultaneously used the reaction control jets of the ballistic control system. When the aircraft passed the 118,000-foot mark, he seemed to have finally regained control; the spin had stopped. The X-15 was now in a 45-degree de-

scent at 4.7 times the speed of sound and was losing 2,650 feet of altitude per second.

However, the aircraft went into a limit cycle with rapid pitching motion. When the X-15 reached 62,300 feet, it disappeared from the radar screens, and telemetry transmission also ended abruptly.

Communications officer Knight tried to obtain information about Adams and his aircraft from an observer aircraft: "Observer 4, do you have information about him?" The pilot replied "no." Knight tried again to make contact with the X-15: "Okay Mike, do you hear me?" Seconds later, instead of Adams the pilot of the observation aircraft broke in: "Pete, I see dust on the salt lake down there."

The X-15 had crashed near the town of Randsburg in California. Pilot Mike Adams was dead. Ten minutes and thirty-five seconds after release from the NB-52, the aircraft had broken up in the air. The aircraft's fuselage had been unable to withstand the increasingly dense atmosphere and the rising aerodynamic loads.

The investigating committee's report into the accident placed the principal blame on the pilot. His instruments had provided him with information, but Adams had nevertheless lost control of the aircraft because he had obviously misinterpreted the information. Perhaps the high spin rate had made him dizzy and he had lost orientation. The additional workload caused by the electrical disturbance at the start of the mission may also have distracted him.

Mike Adams was buried in the city of Monroe, Louisiana. During his tragic mission, he had exceeded the 50-mile-high boundary of space as defined by the Air Force, and therefore he was posthumously awarded the astronaut badge. This made him the first American astronaut to lose his life in a spaceflight.

Firefighters in protective suits extinguish the burning wreckage of the X-15. *NASA*

The rescue team has arrived at the crash site, but all help came too late. *NASA*

Twenty-five years later the name Michael Adams was immortalized at the memorial for astronauts killed in accidents at the Kennedy Space Center in Florida. And about 35 miles northeast of the Edwards Air Force Base, there is a memorial at the crash site dedicated to the memory of the accident.

Project management drew its lessons from the crash, and two important changes were introduced. One concerned the technical aspect of the crash. Since the control center had had no real-time data about the position, attitude, and movement of the aircraft, the flight controllers were unable to help the pilot solve the problem. Indicators were installed to display just such information based on telemetry data from the aircraft.

The other change affected the selection of pilots. Because it could not be excluded that Mike Adams had lost consciousness because

Experts examine the wreckage of the X-15. *NASA*

Commemorative stone at the crash site in the Mojave Desert.

of the rapid rotation of the aircraft, stricter selection criteria were introduced for future X-15 pilots. In addition to the medical examinations already in use, tests were added to determine the candidate's susceptibility to vertigo attacks. This sort of test is now an important component of the suitability tests for future NASA astronauts.

The X-15 program, whose end had more or less been decided, was initially continued somewhat halfheartedly. After eight more flights, in October 1968 it was finally ended.

The X-15 flights had, however, helped solve many scientific and technological questions. The results of the program flowed directly into the Apollo moon-landing program, which was at the verge of reaching its high point. The greatest beneficiary, however, was the space shuttle program, which followed years later. The problems of atmosphere reentry, especially of control at great altitudes, and the effects of extreme heat development caused by air friction were well understood because of the X-15. They flowed directly into the design and construction of the space shuttle fleet.

ODYSSEY IN SPACE: APOLLO 13

A pollo 13 began its fateful journey on April 11, 1970. The clocks at Cape Kennedy showed 1413. In Europe it was already 2013.

The launch was initially trouble free, and the first stage of the mighty Saturn V rocket functioned perfectly. After the stage shut down, it was jettisoned and the second stage ignited. Then, about five minutes after liftoff, there were problems. Two minutes before the anticipated end of burn, the commander reported that the inner of the five engines of the Saturn second stage had shut down. At Mission Control in Houston, a decision was quickly made to carry on. The reduced thrust resulting from the loss of the engine would be compensated for by increasing the burn duration of the remaining four engines by thirty-five seconds. The loss of residual speed would be made up by extending the third-stage burn phase by nine seconds.

Despite the incident, the spacecraft achieved an almost circular earth orbit at an altitude of 115 miles twelve minutes and forty seconds after liftoff. It was determined that the engine shutdown had been caused by severe pogo oscillations after launch. The oscillations had obviously interfered with the flow of fuel to the engines. The control computer subsequently caused the emergency shutdown of the most affected engine. The incident had no additional effects on the planned course of the mission. Apollo 13 would still be able to reach the moon.

The loss of the engine was, however, only the last of an entire series of events that almost prevented the Apollo 13 mission from going ahead. After the launch of the third moon-landing mission had been scheduled for

The launch of Apollo 13 on April 11, 1970.

The original oxygen tank from Apollo 13 at the manufacturer, North American Rockwell, before it was delivered for installation in the service module of Apollo 10. *NASA*

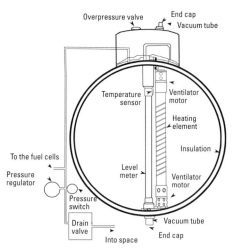

Inner life of an oxygen tank as used in the service module of the Apollo spacecraft. *NASA/Woydt*

April 11, 1970, on March 24, during mission preparations, a problem appeared in the service module.

An oxygen tank that supplied three fuel cells to produce power, rather than provide breathing oxygen for the astronauts, behaved abnormally. Two of these spherical tanks were installed in the supply section of the capsule. In each was a pressure sensor, an electrical heating element, and two stirring fans. These were activated from time to time to stir the contents of the tank, which tended to settle into layers. Each tank also had a safety valve to allow the contents to be drained if necessary.

Both tanks had been filled during a test countdown and were subsequently supposed to be emptied again with the help of the safety valves. On tank no. 1 this worked as envisaged; however, tank no. 2 was only partially emptied. NASA engineers consulted the company that made the oxygen tanks, but a second attempt

also failed. It was then decided to warm the contents of the tank by using the heating element and to use the resulting increased pressure to allow the oxygen to escape through the safety valve. The procedure worked, but it took more than eight hours.

A second cycle of filling and emptying the tanks produced a similar result. Replacing the tanks in question would have been extremely costly, and therefore it was decided to leave them onboard. The refilling procedure had finally functioned normally, and that is what finally mattered. Under normal conditions the tanks would not have to be emptied during the flight.

Despite this irregularity, the launch preparations initially proceeded smoothly. Then, however, something happened that had the potential to throw over the meticulous sequence of preparations. On April 6, just five days before launch, the crew was in the final phase of its training. That day, doctors discovered the highly infectious rubella virus in the blood of Charlie Duke, the backup crew's lunar module pilot. Since the members of the primary and backup crews had been in constant contact during their joint training, all six astronauts were thoroughly examined. Jim Lovell and Fred Haise had already had the childhood illness and were thus immune to another infection, but Ken Mattingly's situation was different. There were no antibodies in his blood, meaning he had never contracted rubella. It was thus entirely possible that he had caught it from Duke, and the doctors feared, not without justification, that he might come down with the disease during the coming flight.

A difficult decision had to be made. One option would have been to exchange the entire crew for its backup crew. This choice was of course eliminated because of Duke's illness. To avoid completely overturning the timetable, the only choice was to replace just Mattingly, with Jack Swigert, the backup crew's command module pilot.

On April 9, with just two days remaining before the launch, Lovell, Haise, and Swigert began training together. Although they could put in only twelve hours together in the simulator, the newly created crew was declared ready to fly.

The mission commander was forty-two-year-old James "Jim" Lovell, a very experienced astronaut. The naval aviator, married and the father of two children, had made his first space-flight in 1965 as pilot of Gemini 7. One year later he was made commander of Gemini 12. The high point of his NASA career to date had been the flight of Apollo 8. In December 1968, Lovell and his two crewmates had become the first men to fly to the moon. They circled earth's satellite ten times on Christmas Eve and then returned safely to the earth.

The original crew of Apollo 13: James "Jim" Lovell, Thomas "Tom" Mattingly, and Fred "Fredo" Haise. *NASA*

Command Module Pilot John "Jack" Swigert. *NASA*

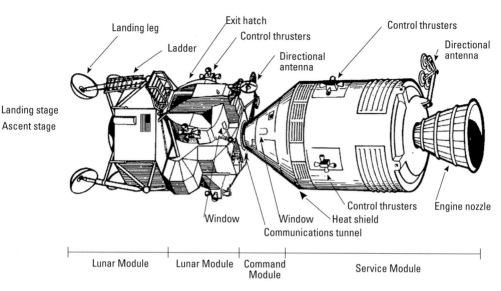

Landing leg — Exit hatch — Control thrusters — Control thrusters

Ladder — Control thrusters — Directional antenna — Directional antenna

Directional antenna

Landing stage
Ascent stage

Window — Window — Heat shield — Control thrusters — Engine nozzle

Communications tunnel

| Lunar Module | Lunar Module | Command Module | Service Module |

The configuration in which Apollo 13 spent most of its time in flight.

Fred "Fredo" Haise, thirty-six years old and the father of four children, was selected to become an astronaut in 1966. The flight of Apollo 13 was the naval aviator's first and, as it turned out, only flight into space.

The new addition to the crew was also a space rookie. John "Jack" Swigert was thirty-four years old and an Air Force pilot. Like Haise, he had joined NASA in 1966. His appointment had caused something of a stir, since he was the first unmarried member of the astronaut corps. The coming lunar mission would be Swigert's only spaceflight.

Apollo 13 was supposed to proceed to the Fra Mauro highlands in the eastern part of the Sea of Storms. The three original crew members, Lovell, Haise, and Mattingly, had chosen the name Odyssey for their Apollo capsule, as a reference to the wanderings of Odysseus, the hero of Greek antiquity. The lunar lander was given the name Aquarius, after the constellation of the Northern Hemisphere.

Despite the engine problems at launch, after one and a half orbits of the earth the third stage of the Saturn V ignited without further ado after all systems had been checked thoroughly. Apollo 13 was on its way to the moon. The command module separated from the rocket stage. Jack Swigert turned the capsule around, docked with the lunar module, and pulled it from its protective shell.

The astronauts and Mission Control all were in high spirits. Jokes and a casual conversational tone characterized the radio traffic. The world public's interest in the mission was, however, rather limited. After the previous two successful moon landings by Apollo 11 and Apollo 12, in the public perception moon landings had obviously become an everyday occurrence. Reports about Apollo 13's flight to the moon were limited to the regular daily news broadcasts.

Forty-six hours after launch, oxygen tank number 2 in the service module, which had previously attracted attention, again made its presence known. Following a routine directive from Mission Control, the crew switched on the ventilators inside the tanks to stir the fuel. As a result, within a few seconds the gas pressure rose to beyond the measuring limits of the sensors; however, no particular importance was attached to it. The technicians in Houston suspected a defect in the wiring of the sensors, but a second attempt resulted in identical behavior.

After two more days in space, Apollo 13 was 205,000 miles from earth. The crew had just ended a television broadcast in which they activated the lunar lander's systems. Fred Haise then shut down the Aquarius's systems again. Commander Lovell was in the process of stowing the television camera. The flight controllers again instructed the crew to activate the tank-stirring fans in the oxygen and hydrogen tanks of the service module. Swigert acknowledged and carried out the directive. Fifty-five hours, fifty-four minutes, and fifty-four seconds after launch, an alarm signal sounded inside the capsule. The cause was a sudden loss of voltage in main power line B. The crew reacted immediately and shut off the alarm. Swigert reported to Mission Control: "Houston, I believe we've had a problem here." The duty capsule communicator, or CAPCOM, responded: "This is Houston, say again please." Commander Lovell repeated: "Houston, we've had a problem. We've had a main B bus undervolt."

At the moment the tank-stirring fans had been switched on, the astronauts had heard a loud bang whose origins could not at first be localized. It was April 13, 1970, and in Houston the clocks showed 2108 Eastern time.

A group of astronauts and controllers watch telemetry data coming in from Apollo 13 during a critical phase of the flight. *NASA*

With the help of instrument readings and incoming telemetry, Mission Control and the crew tried to form a picture of the situation. Then Lovell delivered the next bad news: "We are venting something out into space." What Lovell did not know at that time was that the cloud he saw in front of his window was oxygen. Not only was it needed for the crew to breathe, it was also needed by the fuel cells to produce electricity. Of the three tanks, two stopped working within a few minutes due to the loss of oxygen, and the performance of the third began to falter. Mission Control instructed the astronauts to reduce their oxygen consumption as much as possible and to switch off all systems in the spacecraft that were not necessary for their survival.

A little over an hour after Swigert's first report that something was wrong, a NASA spokesman went before the press. He informed the public about the incident and cautiously advised that the mission could not be carried out as originally envisaged: "Because of the situation that arose this evening, we here in Mission Control are considering an alternate mission around the moon, using the lunar lander's power supply."

Despite the late hour, 2334 Houston time, news of the problems onboard Apollo 13 spread like wildfire. All the major broadcasters in the United States interrupted their regular programming and began special broadcasts. Off-duty mission controllers rushed to help their colleagues. Employees of the companies that manufactured the command and service modules and the lunar module—North American, Rockwell, and Grumman—spontaneously reported to their places of work and offered their help. Astronauts led by Ken Mattingly occupied the spacecraft simulators in Houston and Cape Kennedy and tried to re-create what had happened.

The situation at that time was already assuming dramatic proportions. Fuel cells 1 and 3 were no longer functioning, while indications from oxygen tank 2 showed no pressure, and the pressure in tank 1 was dropping steadily. Gas was still streaming into space from the service module. Obviously it was the contents of tank 1. The cloud around the spacecraft hindered the outside view considerably. This made it more difficult to navigate the spacecraft, since the astronauts needed a clear view of the stars.

Meanwhile, the first immediate step had been to reduce power consumption in the command module to the absolute necessary minimum. The attempt to divert oxygen from the defective tank 1 to the seemingly undamaged tank 2 failed. Slowly but surely it dawned on all those involved: the threat to the mission was the least of their concerns—the lives of the three astronauts on Apollo 13 were now at stake.

The command module's air and energy reserves were calculated exclusively for the few minutes of reentry into earth's atmosphere after the capsule and service module had separated. The majority of the oxygen supply was in the service module, but it had been lost. The power generation system had broken down. At that moment the weak batteries of the command module were the only functioning source of power—apart from the lander's energy supply.

The idea of using Aquarius as a sort of lifeboat seemed to be the only promising way out of the dilemma. Lovell and Haise went into the lunar module and activated its onboard systems. Meanwhile, Swigert switched off all of Odyssey's systems; the command module was practically dead. Finally, Swigert followed his comrades into the lunar module.

But how were they to proceed now? A landing on the moon was of course out of the question. All that mattered now was to get the three men safely back to earth as quickly as possible.

One of the possible scenarios was an immediate return. However, this would require a long burn phase by the service module's propulsion system. This procedure was also impossible because of the added weight of the lunar module. And they really could not do away with Aquarius. Also, the flight controllers feared that the engine might have been damaged by the explosion in the service module. No one wanted to think about the possible consequences of a malfunction. It was therefore decided to let the spacecraft fly around the moon and then back toward the earth in a "free return trajectory." This did, however, require a thirty-one-second burn phase by the lunar module's engine for course correction. The very difficult and time-sensitive maneuver had to be flown manually, but it was successful. Six and a half hours after the explosion, Apollo 13 was on a flight path that would take it around the moon and back to earth.

In Houston the flight controllers realized that a possibly even-longer burn phase would be necessary to hit the landing area on the earth precisely. Without further corrections, Apollo 13 would splash down in the Indian Ocean after 155 hours of flying time. Another possibility was to allow the spacecraft to land in the Atlantic Ocean after just 133 hours. Both possibilities were rejected, however, because the recovery fleet was already on its way to the Pacific. The decision was therefore made in favor of a third variant in which the spacecraft would splash down in the Pacific near the recovery vessels after 143 hours.

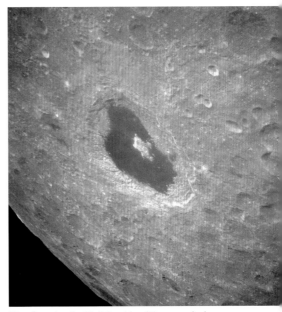

View from Apollo 13 of the side of the moon facing away from earth. *NASA*

At first, however, Apollo 13 moved ever farther away from the safety of home. A little more than seventy-seven hours after launching from the Cape, the badly damaged spacecraft disappeared behind the lunar horizon, and radio communications were lost. Twenty-five minutes later, Apollo 13 emerged from communications blackout on the other side of the earth's satellite. The spacecraft had come to within 158 miles of the moon—and yet it had been unreachable for the astronauts. Despite the seriousness of their situation, they could not ward off the fascination engendered by the view. Excited as children, they described the spectacular landscape that passed beneath them.

However, reality quickly overtook the men again. To save even more energy, the astronauts were forced to reduce the cabin heating to a just-bearable level; it was quite cold onboard. The temperature in the Aquarius fell to 52 de-

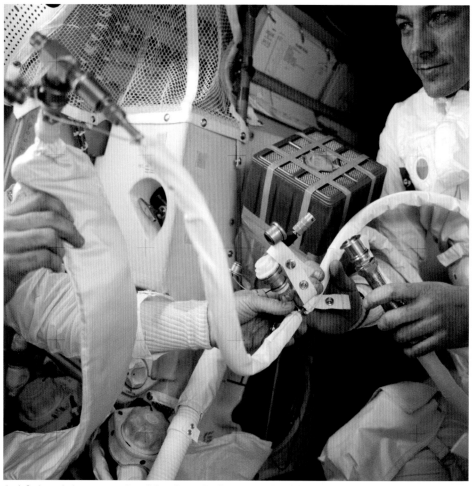

Jack Swigert during construction of the adapter for the command module's lithium-hydroxide canister. *NASA*

grees, while in the shutdown command module it was just 34 degrees. Condensation formed on the instrument panels, and the fogged windows made it almost impossible to see outside. The three astronauts were dressed in just thin flight suits. It was not surprising, therefore, when Fred Haise developed a painful urinary tract infection, with a fever that eventually rose to 102 degrees. His two crewmates treated him as best they could with what was in the medical kit;

nevertheless, the lunar module pilot's physical condition deteriorated visibly.

The men ate their meals cold to avoid consuming more energy by heating them. This and the lack of sleep definitely did nothing to improve the situation onboard, but the three astronauts had no choice but to endure these hardships.

After seventy-nine hours and thirty minutes, Mission Control required the crew to carry out another fine course correction maneuver.

The necessary systems in the lunar module were briefly activated, the ascent engine fired for 272 seconds, and then the navigation and propulsion systems were again turned off.

As time passed excruciatingly slowly, a new potentially life-threatening problem grew steadily worse. The quality of the air the astronauts were breathing was deteriorating as the carbon dioxide they exhaled increased the concentration of it in the atmosphere. If a solution was not found, the men would likely soon suffocate.

Harmful components were usually removed from the air inside a spacecraft by chemical filters. To achieve this, the air was passed through canisters filled with lithium hydroxide. Their capacity was somewhat limited, however, and the cartridges had to be replaced from time to time. Now, however, there were three instead of two persons in the lunar module, and the supply of lithium canisters was quickly running out. Fortunately, there were more of these canisters in the command module, so why not use them? That was easier said than done. The command and lunar modules had been developed and built by different companies, and each had come up with its own design for the system: the fittings of the two canister types did not match. An adapter was required. In Houston, technicians and astronauts worked in all haste to come up with a solution. Using materials available in the spacecraft, they came up with a hose that could be attached to the two different connections. The list of required materials was prepared and assembly instructions were meticulously radioed to the astronauts.

It took longer than half an hour, but finally it was finished. Expectantly the astronauts fitted a canister from the command module to the portable spacesuit life support system. Its ven-

tilators drew in the used cabin air and caused it to flow through the improvised structure. The astronauts soon saw the result they had been hoping for. The concentration of harmful gas dropped from minute to minute until it finally reached a harmless level. For the rest of the flight, carbon dioxide presented no problems.

Meanwhile, the men on the ground took an active part in the dramatic events in space. The media, which a short time ago had treated Apollo 13 so shabbily, pounced on the sensational story. No other events of those days could compare with the disaster in space. Even Pope Paul VI prayed for the three astronauts and for a positive outcome to the rescue effort. The heads of state of many nations, including the Soviet Union and China, offered their communications facilities for radio traffic with Apollo 13. They promised unlimited support in the event of an emergency landing within their

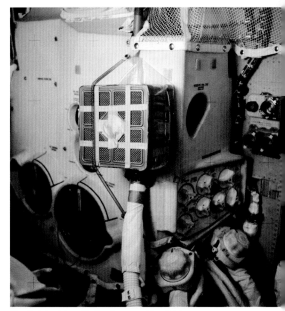

The finished installation for attaching lithium-hydroxide canisters from the command module to the lunar module's life support system. *NASA*

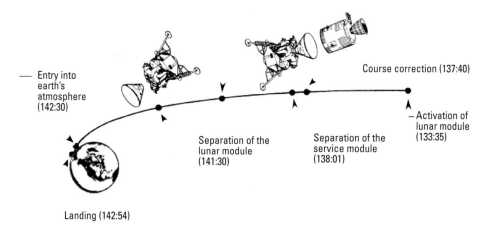

Activation of the command module (140:10)

Entry into earth's atmosphere (142:30)

Course correction (137:40)

Activation of lunar module (133:35)

Separation of the lunar module (141:30)

Separation of the service module (138:01)

Landing (142:54)

Separation of the service module and the lunar module. The figures in brackets refer to the time expired since launch in hours and minutes. *NASA/Woydt*

territories. Russian cosmonauts sent their best wishes and hopes for a successful return by the three Apollo astronauts.

Meanwhile, Apollo 13 was approaching the earth with increasing speed. Mission Control constantly monitored the spacecraft's flight path. They calculated that at least one more correction maneuver would be necessary to reach the target area, and the required maneuver took place 105 hours after the start of the mission. Aquarius's ascent engine operated for fourteen seconds.

Finally the crew began charging the command module batteries, which would be a vital part of reentry. They laid a provisional line through the communications tunnel between the two spacecraft and fed power from the lunar module's batteries to those of the command module. The process lasted sixteen hours, but when it was over the astronauts were sure that they had enough battery capacity for reentry and landing.

At 137 hours and 40 minutes after launch, Odyssey and Aquarius were still 44,115 miles from the earth. They were approaching at 6,800 miles per hour and their speed was increasing. One final ignition of the lunar module's engine placed the spacecraft exactly on course. They were now moving exactly toward the several-mile-wide approach corridor to the earth.

It was now time to jettison the battered service module—Swigert turned the switch. At 138 hours after liftoff from Cape Kennedy, explosive bolts separated the two modules, and the service module slowly moved away from the command module. From the windows the astronauts looked spellbound at the scene before them. Commander Lovell reported excitedly: "And there's one whole side of that spacecraft missing!" And then: "The whole panel is blown out, almost from the base to the engine." The scale of the destruction was even worse than they had feared. The explosion had ripped a large hole in the side of the service module, revealing its interior. Broken wires hung out. The three astronauts quickly took a few photos, and then they turned to more-important tasks.

Four and a half hours before the start of re-entry, the crew began reactivating the command module. This procedure had never been carried out in space conditions before. How would the onboard systems have survived the icy temperatures just above the freezing mark? What about the condensation? Had there been short circuits in the electrical system? Could the onboard computer be restarted? And last but not least: Were the parachutes frozen? So many unanswered questions. But the command module had to function! The only alternative was burning up in earth's atmosphere.

To the astronauts' relief, turning the command module back on went according to plan. The electrical system began working, and the computer came to life. There was one thing left to do: after 141 hours and 30 minutes of flight time, Lovell, Haise, and Swigert separated the lunar module from their Apollo capsule. With

sorrow in their voices, they reported the success of the maneuver to Mission Control. "OK, copy that. Farewell Aquarius and we thank you."

Apollo 13 was now just 11,185 miles from earth. Its speed, then 12,600 miles per hour, was to rise to more than 18,000 miles per hour in the next half hour. Finally, Odyssey struck the outermost layers of the atmosphere. The astronauts could only hope that the heat shield, which had been exposed to the cold of space for days, had survived the explosion undamaged. If it had not, friction would cause Apollo 13 to burn up like a meteor. The astronauts took time to thank all those on the ground who had worked so tirelessly to help them. Commander Lovell radioed: "I know all of us here want to thank all of you guys down there for the very fine job you did." CAP-COM immediately replied: "I'll tell you. We all had a fine time doing it."

Then, as expected, radio communication was lost. The growing friction heat shrouded the capsule in a cloud of plasma that radio waves could not penetrate. As expected, the

The badly damaged service module, photographed from the window of the lunar module. *NASA*

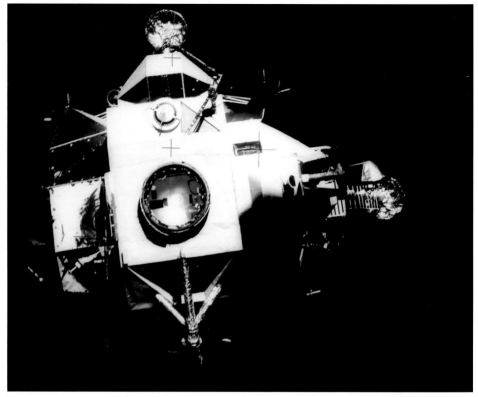

The lunar module has separated and slowly moves away from the command module. *NASA*

"blackout" lasted about four minutes. When that time had expired, Mission Control began calling the spacecraft: "Odyssey, Houston. Come in please." No response. Again and again the call went out: "Odyssey, Houston, over." In Mission Control there was tense anticipation, which gradually turned to disbelief. Had the unthinkable happened? Had the heat shield failed to handle the inferno of reentry? Had it all been in vain? When many had begun to give up hope, a voice was suddenly heard loud and clear from the noise of the receiver. Jack Swigert's brief reply to the ever more urgent calls by CAPCOM Joe Kerwin: "OK Joe." Apollo 13 had survived the final critical phase of the flight.

The blackout had lasted eighty-seven seconds longer than expected. This error was later traced to an arithmetical error. It had simply been forgotten that the capsule was lighter than originally assumed. The weight of the moon rocks that should have been onboard was missing. This had resulted in the spacecraft having a significantly flatter flight path during reentry.

Preparations for splashdown began. Command Pilot Swigert reported the deployment of two drogue chutes. The crew of a helicopter from the recovery ship *Iwo Jima* made visual contact. Moments later, Odyssey could be seen beneath its three red-and-white-striped main parachutes on the television screens in Mission Control. The television picture was seen live by millions of viewers around the world; the number of viewers was surpassed only by the first moon landing.

Apollo 13 splashed down in the waves of the Pacific Ocean southeast of American Samoa five days, twenty-two hours, fifty-four minutes, and fifty-six seconds after lifting off from Florida, just 4 miles from the waiting Iwo Jima. It was April 17, 1970, 1907 Central European time.

Three days and fifteen hours had passed since the explosion, during which the lives of the crew had constantly hung by a thread. The performances of the crew, flight controllers, NASA employees, and the affected companies had been incredible in those eighty-seven hours. The true hero, however, was the lunar module. Aquarius had kept the three astronauts alive under conditions for which the lunar module had never been designed, until the command module was able to cover the last few thousand miles on its own.

The recovery effort in the Pacific followed the familiar routine. Divers jumped into the sea from the helicopter and placed a floating collar around the capsule dancing in the waves. The spacecraft hatch was opened. The three astro-

Moment of great relief: Apollo 13 splashes down in the Pacific on April 17, 1970. *NASA*

US Navy divers help the Apollo 13 astronauts climb out of the command module after splashdown. Jack Swigert is on the left, with his back to the camera. Fred Haise is in the foreground, and Commander Jim Lovell prepares to climb into the life raft. *NASA*

nauts clambered into a life raft, from which they were pulled up to a helicopter. Fred Haise was the first and Jack Swigert followed. Commander Lovell was the last to leave the spacecraft. The recovery action was the fastest to

President Nixon greets the crew of Apollo 13 after their return at Hickam Air Force Base on Hawaii. *NASA*

date for the Apollo program. Less than an hour after splashdown the three men stood on the deck of the *Iwo Jima*. The commanding admiral and the captain of the ship gave speeches, and a chaplain said a prayer. Lovell, Haise, and Swigert were too exhausted for a response.

An initial quick medical examination of the three astronauts showed that they were tired and worn out. Apart from Fred Haise's urinary tract infection, they were in satisfactory condition, and they were immediately flown to Hickam Air Force Base on Hawaii.

There the American president honored them with the Freedom Medal, the United States' highest civilian decoration. In his speech, President Nixon said, "The three astronauts did not reach the moon, but they reached the hearts of millions of people, in America and the whole

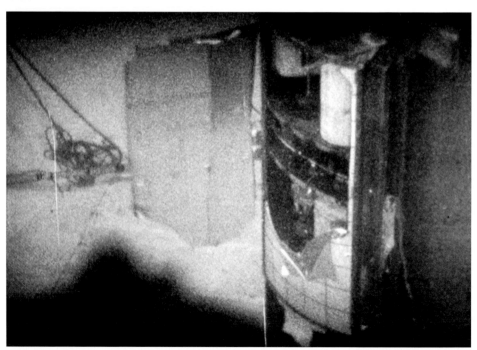

Extract from a film of an experiment conducted by NASA at the Langley Research Center. It shows the moment of the explosion of the oxygen tank, with part of the exterior of the service module flying away. *NASA*

world." During a party, Jim Lovell gave his personal opinion, saying, "We don't realize what we have on earth until we leave it."

The same day that Apollo 13 returned, NASA head Thomas Paine named Edgar M. Cortright, director of the Langley Research Center in Hampton, Virginia, to chair an investigating committee, whose purpose would be to find the cause of the accident and draw the necessary conclusions to prevent a repetition. Named for its chairman, the Cortright Committee had nine members, including Neil Armstrong, commander of Apollo 11.

The committee extensively studied the background of the oxygen tank, its manufacture, and its installation in the service module until the start of the Apollo 13 mission. The committee members reviewed thousands of documents, visited the factories that built the components, and spoke with countless engineers and technicians. After a theory for the cause of the accident had crystallized, a costly experiment was carried out in which a similar oxygen tank was installed in a service module. The theory was confirmed in a vacuum chamber that replicated the conditions in space. The oxygen tank exploded just as Apollo 13's had, part of the service module's shell was blown off, and serious damage was caused in its interior.

The final report by the Cortright Committee arrived at NASA headquarters on June 15, 1970. The report described in detail the course of the accident and its causes. A chain of human errors and technical deficiencies were identified as the cause of the accident. Oxygen tank no. 2 was in fact where the catastrophe had begun. Damaged Teflon insulation in the wiring of the stirring fans had caused a short circuit and subsequent explosion, and the escaping gas had blown away part of the alumi-

num shell. The shock stopped the flow of oxygen to fuel cells 1 and 3, which stopped working within three minutes. The force of the explosion also damaged oxygen tank no. 1 and its lines. For the next 130 minutes its contents escaped into space, after which the service module's oxygen supply was completely expended.

But what had caused the oxygen tank to explode? The cause of the accident went back five years before the launch of Apollo 13. At that time, during the project design phase, the engineers decided to use 65 volts instead of the originally envisioned 28 volts in the spacecraft's power supply. The news of this serious design change obviously failed to reach the manufacturer of the tanks, and they were not adapted accordingly.

It is possible that nothing ever would have happened had the tank not been damaged before the mission. It was originally envisaged for

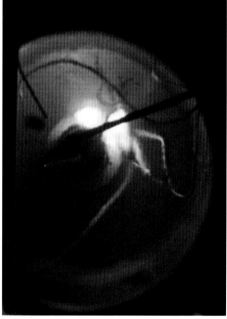

An experiment carried out after the mission shows the burning Teflon insulation inside an oxygen tank. *NASA*

Apollo 10's service module, but due to carelessness during maintenance work, it fell to the ground from a height of two inches. Because of the tank's thin outer skin, this was enough to damage it. Another tank was installed and the damaged one was repaired for use in Apollo 13. Although tests after the repair revealed nothing noticeable, during preparations for the Apollo 13 mission the previously described irregularities occurred during filling and emptying of the tank. By that time at the latest, another detailed examination of the tank should have been carried out. Regrettably, this was not carried out, and it was felt that the problem could be ignored.

And so one thing led to another. A temperature protection switch installed in the tank had never been designed for 65 volts. This protection switch was actuated when the tank was emptied during launch preparations. Because of the high voltage, there was a voltage flashover that resulted in a short circuit. The switch was out of service and had lost its protection ability. The high temperatures created in the process, up to 930 degrees Fahrenheit, damaged the Teflon insulation on a power cable.

When Jack Swigert was instructed to turn on the stirring fan, this caused the cable insulation to catch fire; the fire burned for about thirty seconds. During this time the temperature of the oxygen rose. This caused pressure inside the tank to rise beyond the load limit, and the tank exploded.

From a technical and engineering point of view, the accident was not caused by design errors in the Apollo spacecraft or its components. In a space program of such complexity, there are always unexpected problems that have to be corrected quickly. NASA needed a suitable system for monitoring and implementing design changes. Quality checks on used components had to be improved. In addition, steps had to be taken to ensure that unusual events and anomalous test results were taken note of and resulted in appropriate reactions.

Fortunately the subsequent moon landings were carried out without serious incident.

Incidentally: Ken Mattingly never came down with measles.

The command module is now on display in the Kansas Cosmosphere and Space Center.

Sequence of Events

Date	Time (Cape Kennedy Local Time)
April 11	
13:13:00	Launch of Apollo 13.
13:15:43	First-stage burnout.
13:15:46	Second-stage ignition.
13:16:21	Rescue tower jettisoned.
13:18:30	Premature shutdown of the inner of the total of five second-stage engines.
13:22:52	Second-stage burnout.
13:22:56	First ignition of the third stage.
13:25:39	Entry into orbit around the earth.
15:48:46	Second ignition of the third stage for insertion into the transfer trajectory to the moon.
15:54:37	End of the third-stage second-burn phase.
15:54:47	Transfer trajectory to the moon achieved.
17:14:00	Lunar module separated from the third stage.
April 12	
19:53:49	First course correction; from now on, Apollo 13 is no longer on a free return trajectory.
April 13	
11:15	Crew routinely engages the fuel-stirring fan in oxygen tank 2; contents gauge shows full (obviously an instrument error).
19:33	Commander and lunar module pilot climb into the lunar module and activate its systems.
20:43	Commander and lunar module pilot return to the command module.
21:06:20	Fuel-stirring fan in oxygen tank 2 switched on again; voltage spike in the Apollo capsule's power supply system.
21:06:22	Short circuit inside oxygen tank 2.
21:07:31	Temperature inside oxygen tank 2 rises rapidly.
21:07:52	Temperature reading inside oxygen tank 2 goes to zero (indication of a defective sensor).
21:07:53	Sudden movement of spacecraft about all three axes.
21:07:53	Alarm: undervoltage in main line B; crew hears a loud bang.
21:08:20	Command module pilot reports the incident to ground control in Houston.
21:10:45	Fuel cell 3 fails.
21:22:07	Commander reports that something is flowing into space.
21:22:58	Fuel cell 1 lost.
22:56	Commander and lunar module pilot climb into lunar module.
23:13	Fuel cell 2 is switched off as a safety measure.
23:23	Command module is deactivated at the same time that the lunar module's systems are activated.
April 14	
02:42:43	Second course correction; Apollo 13 is again on a free return trajectory to earth.
18:21:35	Apollo 13 disappears behind the moon.
18:46:10	Apollo 13 emerges from behind the moon.
20:40:39	Insertion into transfer trajectory to earth.
April 15	
22:31:28	Third course correction.
April 17	
06:52:51	Fourth course correction.
07:14:48	Service module is separated.
09:23	Command module is activated.
10:43:00	Lunar module is jettisoned.
11:53:45	Reentry into earth's atmosphere begins.
12:01	Contact with Apollo 13 reestablished after radio interruption.
12:07:41	Splashdown.
12:32	Command module hatch opened; crew leaves the capsule.
12:42	Crew onboard recovery helicopter.
12:53	Crew onboard recovery ship.
13:36	Command module on recovery ship.
April 18	
12:20	Crew leaves the recovery ship for Hawaii.

DEATH BY SUFFOCATION: SOYUZ 11

The Soviet Union had lost the race to the moon. In July 1969, the United States had succeeded in achieving what the Soviets would themselves have liked to ascribe to their banner, in order to demonstrate the superiority of their system to the hated capitalist West. But the spectacular first landing by Apollo 11 was quickly followed by other lunar missions. And NASA had come out of the recent Apollo 13 mission, which had almost ended in failure, even stronger than before.

The Soviet Union looked for new missions and goals. One possibility was a permanent manned space station. Since 1964, the Soviets had been working on a military space station code-named Almas (Diamond). Because the work had not proceeded as quickly as hoped, in 1969 it was decided to work on a civil variant as well. On April 19, 1971, the first of these stations, named Salyut (Salute), was launched.

Three days later the spacecraft Soyuz 10 followed with a crew of three to place the station into operation in orbit. The launch was a failure, however. A few seconds before engine ignition, one of the launch complex's masts could not be retracted. To carry on with the launch would have been too dangerous—contact between the rocket and the obstacle would undoubtedly have had disastrous consequences. The launch was called off for that day, and a new attempt was scheduled for the next day. But the same thing happened once again, and it was decided to accept the risk and launch anyway. Fortunately, all went well and the three cosmonauts, Vladimir Shatalov, Alexei Yeliseev, and Nikolai Rukavisnikov, were on their way to space.

The next day, April 24, 1971, Soyuz 10 docked with the space station. Although the docking maneuver had been carried out successfully, not all of the electrical connections between the two spacecraft could be established. The Soyuz capsule's computer recognized this and reacted by igniting the Soyuz thrusters for thirty seconds, and the spacecraft turned 30 degrees around the docking adapter. The Soyuz's docking cone was damaged, preventing an airtight connection from being established. The crew could not reach the space station by carrying out a spacewalk, because for reasons of a lack of space and weight considerations the cosmonauts had no spacesuits with them. There was no choice but to abort the mission. Uncoupling the Soyuz proved difficult. After they finally succeeded in separating from the station, the three cosmonauts returned to earth empty-handed.

As always during the Cold War, the Soviet state's news agency, TASS, reported that the mission had been a complete success. Not a word mentioned that the cosmonauts were supposed to have boarded the space station.

The failure of the docking maneuver was attributed to problems with Soyuz 10's active docking adapter. The Soyuz docking adapter was obviously damaged in the forceful coming together of the two spacecraft. However, there was justified hope that its opposite number on the Salyut space station had survived undamaged.

In mid-May that year, Soyuz 11, with an improved docking mechanism, was declared ready to launch. Its crew consisted of the experienced cosmonaut Alexei Leonov, the first man to walk in space, and two space rookies, Valery

Kubasov and Pyotr Kolodin. A few days before the names of the cosmonauts were announced, Kubasov received bad news: X-rays of his lungs had revealed a dark spot. Since this was a possible indication of tuberculosis, Kubasov could not possibly fly; the rules for such a case were clear. It was not permissible to replace a single member of a crew, so instead the entire crew had

The crew of Soyuz 11 during training in the simulator. From left to right: Victor Patsayev, Georgi Dobrovolsky, and Vladislav Volkhov. *RKK Energia*

to be replaced by a backup crew, which consisted of thirty-four-year-old commander Georgi Dobrovolsky, thirty-seven-year-old Viktor Patsayev, and thirty-five-year-old Vladislav Volkhov. Soviet air force pilot Dobrovolsky had been a member of the cosmonaut corps since 1963. He had initially trained for the Russian lunar program, and after it was canceled he was trained for a mission on the space station. Patsayev, who had been a cosmonaut since 1968, had worked since 1958 on the development of space vehicles in a design bureau. For both men the flight on Soyuz 11 would be their first space mission. Vladimir Volkhov, however, had twice been to space, most recently on the Soyuz 7 mission in October 1969. Unlike his colleagues, he had no flying background, but despite this he was accepted into the first group of Soviet cosmonauts.

Leonov and Kolodin, members of the original crew, protested vigorously against the complete replacement of the crew. At first they even found support for their suggestion that only their comrade Kubasov be replaced by Volkhov. On June 4, just two days before

launch, a committee of high-ranking representatives from politics, the military, and the space program issued the decree. The entire crew would be replaced. Dobrovolsky, Patsayev, and Volkhov would go into space aboard Soyuz 11.

The launch on June 6, 1971, was trouble free; the booster rocket carrying Soyuz 11 lifted off at 0755 Moscow time and nine minutes later entered orbit. The next day, after two course corrections and an automatically controlled approach, the docking maneuver was carried out with the Salyut space station. This time everything worked, and a hard dock was achieved. After four orbits by the combined spacecraft, Viktor Patsayev transferred to the Salyut and became the first crew member of a space station in the history of spaceflight. However, because of the smell of burning plastic in the space station's atmosphere, the three cosmonauts spent another night in the Soyuz capsule. The next day the smell had disappeared. The three men could finally begin putting the space station into operation.

This is the last, rather blurry, photo taken by the crew of Soyuz 11 after undocking from the Salyut space station. *RKK Energia*

Salyut

The moon landings of the Apollo era marked a turning point for the Soviet space program. In the Soviet Union the influential leadership circle continued to press for the continuation of the Soviet moon-landing program. At the same time, realistic-thinking representatives from the space industry sought new objectives that would be more promising, as they seemed achievable. A group of them proposed the idea of launching an initiative, the centerpiece of which was to be a space station. It could be achieved quickly and cost-effectively on the basis of existing technology. The project was to build on a military station called Almaz (Diamond), which had been under development since the mid-1960s. The project had been given the highest priority when it became obvious that the Soviet lunar program was doomed to failure. Despite this, work on the Almaz project had proceeded slowly. Beginning in 1969, however, work began in parallel on a civilian project designated DOS 7K.

The basic framework of the Almaz station was used for this project, and proven technology from the Soyuz spacecraft, adapted for long-term use, was installed. The first station was completed in record time. Called Salyut (Salute), it was launched on October 11, 1971.

A total of nine Salyut space stations were launched. Only six of these were ever manned, however. Altogether, sixty-five cosmonauts spent up to 237 days on the various stations.

Name	Launch	Days in Space	Days Manned	Use
Salyut 1	April 19, 1971	175	24	civil
Salyut 2*	April 4, 1973	54	–	military
Salyut 3*	June 25, 1974	213	16	military
Salyut 4	December 26, 1974	770	93	civil
Salyut 5	June 22, 1976	412	67	military
Salyut 6	September 29, 1977	1,764	683	civil
Salyut 7	April 19, 1982	3,216	816	civil

*Military Almaz station, given the civilian designation Salyut only as a cover.

The Salyut stations were each about 49 feet long, with a volume of about 3,500 cubic feet and a weight of 20.4 tons. Energy was supplied by a total of four solar panels with a combined area of 550 square feet. Including these panels, the station had a span of 55.75 feet. In the rear part was a propulsion unit, which, like the solar panels, had been taken from the Soyuz space capsules almost unchanged.

On the front end was the docking mechanism with airlock, through which the crews delivered by Soyuz capsules entered and exited the station. Salyut 7 and Salyut 8 were modernized versions of the original type; they had two docking cones instead of one. This allowed the crews to be supplied by unmanned Progress space freighters even if their Soyuz capsule was occupying one of the cones.

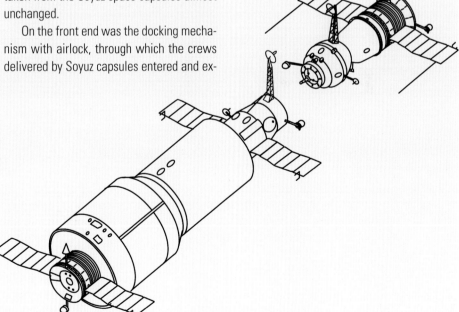

Schematic drawing of the Salyut 1 space station with a Soyuz space capsule just prior to docking. *NASA*

Soyuz 11 on the launchpad.

The launch of Soyuz 11.

For Dobrovolsky, Patsayev, and Volkhov, there began an extensive program of scientific experiments from the fields of medicine, biology, physics, and astronomy. The men on the station were doing well. Over the next days and weeks, in television broadcasts they showed themselves to be in good spirits and relaxed, revealing a very different side than had been expected in the West—the Russian cosmonauts had earned the reputation of being cool, emotionless robots without feelings. The three cosmonauts became true heroes of the people in the Soviet Union. The Soviet media reported on the men and their work in orbit in detail that was unusual for that time. For the first time in many years the space program of the Soviet Union had been able to achieve a real success and a first in space travel.

On the evening of July 29, the three men left the Salyut space station and returned to their capsule to begin the return to earth. They sealed the tunnel linking the spacecraft and the space station, entered the Soyuz landing module, and closed the hatch between it and the orbital module. As they were doing this, a warning light came on, signaling that the hatch had not closed correctly. Ground control directed the three cosmonauts to repeat the procedure, but there was no change and the warning light remained on. The search for the cause revealed that an electrical contact was obviously not functioning correctly. Therefore, the "hatch closed" signal could not be sent. Dobrovolsky was instructed to fix the contact with a piece of insulating tape. This had the desired effect, and the warning light went out. In addition, pressure tests were carried out, and these were all found satisfactory by the crew and control station. Apparently nothing stood in the way of their return to earth.

Late on the evening of June 29, 1971, the space capsule separated from Salyut 1. Before leaving, Soyuz 11 circled the station several times. Patsayev took a series of photos to document the condition of the space station. The control center wished the crew good luck, then radio contact with the spacecraft was lost.

At 0135:24 Moscow time, the automatically controlled return to earth began. The braking thrusters fired for 187 seconds as the spacecraft flew over the Pacific Ocean. Twelve minutes later the planned separation of the orbital and landing modules took place. When Soyuz 11 came into radio range again at 0149, mission control was unable to establish contact with the cosmonauts. The spacecraft did not respond to radio calls.

Search and recovery teams were sent to the probable landing site. Radar tracking by the Soviet air force picked up the spacecraft exact-

Vladislav Volkhov and Georgy Dobrovolsky during an experiment onboard the space station. *RKK Energia*

Recovery helicopters landed near Soyuz 11.

ly at the predicted time over the Black Sea. A short time later an Ilyushin Il-14 search aircraft spotted the capsule beneath its parachute. The aircraft's crew was also unable to make contact with the three cosmonauts.

Soyuz 11 finally landed at 0217 Moscow time on June 30, 1971, just 6 miles from the predicted site, about 120 miles east of the Kazakh town of Shesqasghan. The mission had lasted twenty-three days, eighteen hours, twenty-one minutes, and forty-three seconds.

The first helicopter arrived a few minutes after the capsule touched down. Its crew rushed to the capsule, which externally appeared to be intact. The members of the recovery team banged on the outside of its steel hull but received no response from inside. When the helicopter crewmen opened the hatch, to

Medics work on the cosmonauts, but they were beyond help.

their dismay they found the cosmonauts lying motionless in their seats. At first glance they appeared unhurt. On closer examination, however, there were dark blue spots visible on their skin and thin trickles of blood from their noses and ears. The recovery team rushed to get the cosmonauts out of the Soyuz capsule. This proved difficult, however, since the spacecraft was lying on its side, although this was not out of the ordinary.

Eventually all three cosmonauts were removed and resuscitation efforts began. These were unsuccessful, however. The three cosmonauts were dead.

After a cursory examination, medical staff on the scene diagnosed hemorrhaging in the lungs and brain. Also, the nitrogen content in the blood was very low, all of which were clear

indications of death by suffocation. An inspection of the interior of the spacecraft revealed nothing unusual. Everything was as expected, except for the fact that one of the two valves of the life support systems was open. This aroused suspicion that the cabin's atmospheric pressure could not have been at the expected level.

In reaction to the first reports about the fatal outcome of what had been such a successful mission, a public collective grief without parallel in the history of the Soviet Union seized the population. Radio and television broadcasters interrupted their programming and reported on the tragedy that had claimed the lives of three people's heroes. Heads of state from all over the world sent condolences.

The physical remains of the three cosmonauts were taken to Moscow and laid out in the Central House of the Soviet army amid a sea of flowers. Thousands of ordinary Soviet citizens and many high-ranking representatives of state and party, including Leonid Brezhnev, paid their last respects.

At the burial of Vladimir Komarov four years earlier, an offer by the United States to send an official representative to take part in the funeral had been rejected by the Soviets. The reason given was that it was a "private matter." This time the Kremlin authorities accepted the American suggestion. Tom Stafford, a NASA astronaut, took part in the state funeral, which was broadcast live on television. The three cosmonauts had been cremated, and Stafford even served as one of the three bearers of the urns. Dobrovolsky, Patsayev, and Volkhov were interred in the Kremlin Wall Necropolis in Red Square.

Autopsies on the bodies at the Burdenko Military Hospital in Moscow had confirmed the initial examination made on-site immediately after the landing: hemorrhaging of the blood vessels in the brain, with lesser amounts of bleeding under their skin, in the inner ear, and in the nasal cavity, were found in all three cosmonauts. Their eardrums had burst and no nitrogen, oxygen, or carbon dioxide was present in their blood. The exposure to a vacuum environment caused the oxygen and nitrogen in their bloodstreams to bubble, and these bubbles traveled to the heart and lungs and had caused embolisms there. Their blood was also found to contain heavy concentrations of lactic acid, a clear indication of oxygen starvation.

There was therefore no doubt as to the cause of death: death by suffocation caused by the sudden loss of the cabin atmosphere. As tragic as the deaths of the three cosmonauts were, those responsible in the East and the West were relieved that the cause for the

Interment of the urns containing the remains of the astronauts in the Kremlin Wall.

deaths of the space travelers was not to be found in the time they had spent in space. The crew of Soyuz 11 had spent almost twenty-three days in space, five days longer than the previous record set by Soyuz 9 and nine days longer than the then-longest American mission by Gemini 7.

Although no telemetry had been received from Soyuz 11 during the descent phase, the state investigating committee was at least able to investigate the period in question by using the spacecraft's flight data recorder. The data showed that the landing and orbital modules had separated at an altitude of 90 miles, as planned. Shortly afterward there had been a rapid pressure loss in the cabin atmosphere. The speed of the pressure loss could easily be reconciled with the presence of an open ventilation valve, of which there were two in the Soyuz spacecraft, but they had been closed throughout the entire mission. At least one was supposed to be opened after landing, to equalize the pressure difference between the cabin atmosphere and the outside air and to let in fresh air. Obviously in this case something had gone wrong. Separation of the orbital and landing modules was accomplished by two explosive bolts. Apparently the bolts had not exploded one after another as planned, instead going off simultaneously. This produced mechanical forces that had caused the opening of one of the two valves. The cabin atmosphere escaped into space with a whistling sound. The cosmonauts must have immediately recognized this sound. Dobrovolsky's pulse rate rose from a quiet 80 beats per minute to well in excess of 100 beats in a matter of seconds, while Volkhov's rose to more than 180 beats per minute. The cosmonauts had probably realized immediately that their spacecraft was leaking atmo-

sphere. Because of the previous problems with the hatch, they had probably thought it to be the cause of the pressure loss.

Dobrovolsky obviously released his harness quickly and examined the entry hatch, but then he probably had to realize that it was in order. To locate the source of the whistling sound, the crew had obviously turned off the radio equipment to avoid being distracted by its noise. The three men had a total of only about ten to fifteen seconds for these actions before a lack of oxygen rendered them unconscious. That was far too-little time to be able to localize, let alone repair, the problem. Within one minute and fifty-two seconds of the start of decompression, cabin pressure fell to zero. The cosmonauts were suffocated.

In the course of the accident investigation, Alexey Leonov tried to replicate the accident in the simulator. It took the experienced cosmonaut precisely fifty-two seconds to close the valve in question, far more time than the men on Soyuz 11 had available, even if they had identified it as the cause.

After the total loss of atmosphere, Dobrovolsky, Patsayev, and Volkhov were exposed to the vacuum of space for more than eleven minutes before the recovery team reached them. For this reason there was no chance of reviving the men. Of course, on the ground no one knew this, and the members of the recovery therefore did everything they could to revive the cosmonauts.

A state investigating committee was established and began its work on August 17, eventually making a series of recommendations. For example, it recommended that the ventilation valves be redesigned to make them more resistant to shocks. Another recommendation concerned the installation of easy-to-operate handles for rapid manual operation of the valves.

The most important point in the concluding report concerned the demand that spacesuits be provided for the entire crew. They were to be worn at least during critical situations during which a loss of pressure might be expected, meaning during launch and reentry and all docking and separation maneuvers. What had been a matter of course for the first cosmonauts—namely, wearing such suits in the Vostok spacecraft—had ceased to be standard practice since the flights by the Voshkod capsules. The official version was that spacesuits were no longer required; the Soviet space vehicles were safe to the degree that they were not required. The truth was, however, that a cosmonaut in a spacesuit obviously took up considerably more space than one without one. This had been the only way that a crew of three could be accommodated in a Voshkod spacecraft and later in a Soyuz capsule. Because the Soviet Union was always trying to trump the Americans with new first accomplishments in space, the first launch of a spacecraft with a crew of three had been given a high priority by those responsible. The leadership of the Soviet space program saw itself under pressure from the party and state to ignore safety aspects and to dispense with spacesuits. The cosmonauts had accepted this reluctantly, although many of them more or less openly expressed a different opinion. From then on, spacesuits were found on a Russian spacecraft only when they were needed for extravehicular activities.

After the deaths of Georgi Dobrovolsky, Viktor Patsayev, and Vladislav Volkhov, the Soviet

Memorial at the place where Soyuz 11 landed.

space program initially came to a halt until a more modern spacesuit called the Sokol (Falcon) was available. It was based on the design of an existing suit for combat pilots. The Sokol spacesuit has been used since the resumption of spaceflights with the Soyuz 12 mission in September 1973. The original Sokol K (K for Kosmos) was considerably improved in the years that followed. Now designated Sokol KV2, it is still used today by supply flights to the ISS (International Space Station).

SEPARATION DIFFICULTIES: SOYUZ 18-1

The Soviet Union had tested the first example of a rocket type with the designation R7 in 1957. It had been under development since 1954 for military use as a ballistic missile. After the first three test flights failed, during its fourth test flight on August 21, 1957, it covered a distance of 3,975 miles. The R7 was thus the world's first functioning intercontinental ballistic missile.

Two further test flights, on October 4 and November 3 of the same year, each by a slightly modified variant, went into the history books. The flight on October 4 delivered the first artificial satellite, Sputnik 1, into orbit. The first satellite weighed just 190 pounds, but the flight in November put the much-heavier Sputnik 2, weighing 1,120 pounds, into earth orbit.

The Soyuz booster rocket, named after its main payload, was developed from the early R7 via several intermediate stages. The Soyuz was developed according to the principle of clustering several engines of the same type. Four identical booster rockets were grouped around a central stage. They functioned for 118 seconds and were then jettisoned. The central stage, which also ignited at launch, operated for a total of 292 seconds before its fuel was consumed. The clustering of a total of five propulsion units had several advantages. For one, it made it possible to use simple rocket engines that were easy to control. For another, it had the advantage of making the launch approval dependent on the proper functioning of all of the propulsion units. Should one or several of these malfunction, the launch sequence could be stopped before liftoff, and the rockets remained on the ground.

When the five engines of the first stage had burned out, the second took over, its single engine providing thrust for another 246 seconds. The central stage and the second stage were connected to each other by a grid-like structure and on stage separation were separated by explosive charges.

The Soyuz booster rocket had an overall length of 149.60 feet. The central stage accounted for 91.20 feet of this. At its base it measured 33.8 feet in diameter, and it was capable of putting payloads weighing up to 14,220 pounds into a 120-mile-high earth orbit. All the engines of the central stage, the boosters, and the second stage used kerosene and liquid oxygen as fuel.

At the beginning of the 1970s, preparations for an ambitious project were fully underway: the docking between a Soviet Soyuz spacecraft and an American Apollo capsule. The so-called Apollo-Soyuz Test Project had been launched in 1970, and its successful completion in 1975 contributed significantly to the end of the space race and to understanding between the two superpowers, the United States and the Soviet Union.

In parallel with this historic and, most significantly, politically important plan, the Soviet Union was working hard on its Salyut space station program. The station Salyut 4 was already in space. Two cosmonauts had spent what was, by current standards, a record-breaking time of thirty days on the station. Now, for the second time a spaceship crew was to fly to the station.

This was to be done by the crew of Soyuz 18, which the plan envisaged spending sixty days in space. If the launch took place on the

Soyuz Launch Abort

The launch of a Soyuz booster rocket with the spacecraft of the same name can be broken down into three phases. In each of these three phases there are different ways of getting the space capsule and its crew of three out of the danger area if irregularities occur during or after the launch.

The first phase begins prior to liftoff and includes the first 160 seconds of flight. During this time the cosmonauts have the rescue rockets with which to escape. The towerlike rescue rockets at the tip of the Soyuz capsule are armed sixteen minutes prior to launch. They use several maintenance-free and powerful solid-fuel rocket engines. In the event of danger they are ignited and catapult the orbital and landing modules, which are still covered by a protective shell (1). Moments later, latticelike air brakes deploy, stabilizing the capsule and reducing the extreme acceleration forces. Nevertheless, during this process the cosmonauts can be exposed to forces equal to twenty-one times earth's gravity.

After the rescue rocket's solid-fuel rocket engines burn out, the now-useless tower is jettisoned and the landing module separates from the orbital module and protective shell (2). Carried by its own momentum, it follows a parabolic path. After passing its apex the parachute is released (3). The landing capsule then makes a landing, which largely resembles a landing made from orbit (4).

If the first 160 seconds of flight pass without incident, phase II of the ascent begins. The protective hull and rescue rockets are now of no further use to the spacecraft and are jettisoned (5). In the following 126 seconds, in the event of danger the orbital and landing modules separate from the instrument section and the booster rocket. Moments later the landing module, carrying the cosmonauts, and the orbital module separate (6). Propelled by its own momentum, the landing module returns to earth in a free flight path (7). The not entirely uncritical separation of the first and second rocket stages takes place in this second phase of the ascent (8).

At 522 seconds after liftoff, phase II comes to an end and phase III begins. By this time the spacecraft has almost reached earth orbit. Should a serious malfunction now occur, the Soyuz spacecraft independently initiates a landing with the help of the instrument module's braking thrusters (9). In this case the descent and landing differ little from those at the end of a spaceflight. After the braking maneuver the three components of the Soyuz capsule separate from one another (10). The instrument and orbital modules fall unchecked to earth, while the landing module descends to earth more or less gently.

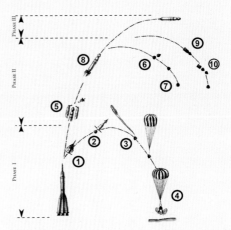

The various rescue options during a Soyuz launch (see text). *RKK Energia*

Commander Vassily Lazarev (right) and flight engineer Oleg Makarov. *RKK Energia*

Lazarev and Makarov shortly before boarding their spacecraft. *RKK Energia*

A rocket of this type launched on April 5, 1975, carrying Soyuz 18. *NASA*

The commander of Soyuz 18 was Vasiliy Lasarev, a doctor and air force test pilot. The forty-seven-year-old cosmonaut from Siberia was married with one child. His flight engineer was civilian Oleg Makarov, born in the central Russian Republic, who was forty-two years old, married, and the father of two children. Lasarev and Makarov had spent two days in space together aboard Soyuz 12 in 1973. This would be the second space mission for both men.

At the Cosmodrome in Baikonur, Kazakhstan, the two men boarded their spacecraft at noon on April 5, 1975. They then took their places, with Makarov in the right seat and Lasarev in the left one. The hatch was closed. The final two seconds of the countdown began.

planned date of April 5, 1975, the cosmonauts would return in June, immediately prior to the launch of Soyuz 19 on the historic double mission by the United States and the Soviet Union, which was planned for July of the same year.

At 1403 Moscow time the engines of the first stage of the booster came to life. After the launch computer had verified that all of the individual engines were functioning perfectly, the restraint clamps that kept the rocket on the ground were released and the Soyuz lifted off. After just sixty seconds it had reached a speed of 930 miles per hour; after 120 seconds, 3,350 miles per hour.

Within 118 seconds the four first-stage booster rockets had used up their fuel and were jettisoned. After 160 seconds of flight without incident, the escape rockets and the protective shell around the capsule were jettisoned. The crew had survived the critical initial-launch phase.

The central stage continued operating until its fuel was consumed. It was time for separation of the central stage and second stage. To accomplish this, two separate systems had to separate the latticelike connection between the central stage and second stage, with the help of twelve explosive bolts. The charges were installed in the lattice structure—six above and six below. The correct functioning of the six lower explosive charges was controlled by an installation in the central stage, while a similar installation in the second stage controlled the functioning of the six upper explosive charges. The two systems were connected by cables. A few seconds after the completion of separation, the second-stage engine was supposed to fire and continue the flight. At least in theory.

This time, everything happened differently. Half the explosive charges on the central-stage side had obviously exploded several seconds earlier than expected. This had been insufficient to separate the two rocket stages from one another, but the electrical link between the two control units had obviously been destroyed. The reason for the malfunction appeared to have been heavy vibration during the ascent, which had caused a relay to switch through. As a result of the line break, it was impossible to send the trigger impulse, which was given at a precise time, to the three explosive charges in the central stage that had not exploded and the six in the second stage.

When the second-stage engine finally ignited, the two rocket stages were still connected by the lattice structure. The hot exhaust gases from the rocket engine melted the connecting structure, and the central stage dropped away, but the rocket with the Soyuz spacecraft had already deviated from the planned flight path because of the diversion of the thrust stream. The rocket's course was no longer directly toward space, which was of course recognized by the rocket's computer. When, after a few seconds, the deviation exceeded 10 degrees from the previously calculated direction of flight, it automatically began aborting the flight. At an altitude of 119 miles it initiated the rescue se-

The lattice structure between the central stage (right) and the second stage (left) shortly before the two components were joined together. *NASA*

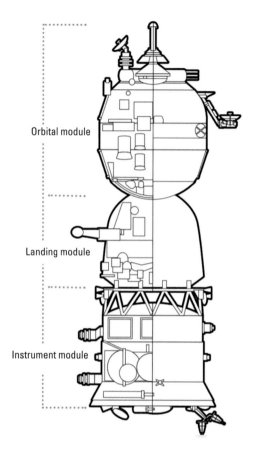

Orbital module

Landing module

Instrument module

The components of a Soyuz space capsule.

quence by issuing the order for the Soyuz capsule to separate from the second stage.

Orbital and landing modules moved away from the upper stage, which was still operating. At a safe distance the Soyuz capsule's onboard computer began the automatic landing sequence and separated the two modules. Because of the extreme flight situation, the cosmonauts were exposed to forces up to twenty-one times their body weight for many long seconds. Amazingly, however, both remained conscious. The Soyuz capsule was now falling back to earth.

The automatic system finally deployed the drogue chute, which a short time later pulled the main parachute from its container. The cosmonauts had survived the worst. After they realized what had happened, the two cosmonauts took matters into their own hands. The first thing they did was to ask where they were going to come down. The two were concerned that their flight path might take them into nearby China, which was not exactly a friendly nation. The flight controllers were able to reassure them, however, and Soyuz 18 finally came down in the western part of Siberia, in a snow-covered, inhospitable, and unpopulated area, more than 480 miles from the Chinese-Russian border and 980 miles from the launch site in Baikonur. Originally planned for sixty days, the flight had ultimately ended after just twenty-one minutes and twenty-seven seconds.

The cosmonauts' difficulties did not end when they touched down, however. The space capsule had come down on a steep slope, and the egg-shaped landing module rolled down it. The movement did not stop until the parachute lines became tangled in trees and bushes. The capsule fortunately came to a stop a few yards from a steep drop-off.

Lasarev and Makarov worked their way out of the capsule. Although radar had tracked their flight path and they were now transmitting homing signals, they had to prepare for a cold night in their secluded landing site. The temperature was 19 degrees Fahrenheit. The cosmonauts lit a campfire and waited for the rescue team, which in fact would not arrive until the next morning.

Usually every effort would be made to keep the incident secret from the public. Because of the Soviet Union's restrictive information policy during the Cold War, it never reported failures, disasters, and accidents—if it could avoid it—which might damage the Soviet Union in world public opinion. Soyuz 18's launch accident, however, happened at a time marked by preparations for the joint space mission with the United States. The Soviet Union therefore saw itself forced to inform its American partner about the incident; however, this did not happen until two days later, on April 7.

The report received by NASA of course immediately awakened doubts about the reliability of the Russian technology. The Soviets quickly gave assurances that it had been an older-model rocket. The one earmarked for the joint project was a new type, and therefore there was no possibility of a repetition. Fortunately, the Soviet optimism proved to be justified. The joint Apollo-Soyuz mission was a success with no noteworthy complications.

In Soviet parlance and until the present day, the failed flight is referred to euphemistically as the "5th of April anomaly." In the West the name "Soyuz 18-1," sometimes called Soyuz 18A, has become accepted. Because of the secretiveness of the Soviet Union during the Cold War period, many legends have grown up around events of that period. Such spectacular events as the emergency landing by Soyuz 18, in particular, fed the rumor mill. For example, some sources claimed that the cosmonauts came down in Chinese territory and were extracted by helicopter in a secret commando-style action without the knowledge of the Chinese government. One can only speculate as to the veracity of such information. The above-described sequence of events does however seem plausible, and this description of events is found in a number of media. It can therefore be accepted with some degree of certainty that the events in April 1975 took place as described. Perhaps one day there will be absolute certainty if Russia decides to open the Soviet secret archives and make them accessible without restriction for publication in the East and the West.

Launch from Baikonur and landing near the town of Gorno-Altaisk at the edge of the Altai Mountains.

JUST MADE IT ONCE AGAIN: SOYUZ T-10-1

In the late 1950s, the development of rocketry was still in its infancy. One launch after another failed. Many rockets exploded on the launchpad, while others blew up immediately after liftoff or in the first seconds or minutes of

The rescue rockets of an American Mercury at the moment of ignition. In this case, however, the system failed and the capsule remained on the rocket—only the escape tower blasted away.

flight. An all-consuming fireball was the result when vast quantities of highly explosive fuel ignited. In addition to the total loss of the booster rocket, failed launches regularly inflicted damage on launch facilities and the surrounding areas. Despite this, plans were developed both in the Soviet Union and the United States to send a man into space as soon as possible, have him circle the earth once or twice, and then land him safely again.

In the United States, NASA engineers, early on, turned their thoughts to how to provide the greatest possible safety to their astronauts during various phases of their flight. Particular attention was paid to the extremely critical launch phase and the first seconds of the ascent.

Maxim "Max" Faget (1921–2004), an American spaceflight engineer, played a major role in the development of the first manned American spacecraft, the Mercury capsule. He was one of the first to rack his brain as to how astronauts might escape the danger zone of an exploding rocket. Faget, therefore, at least in the West, is considered the father of the launch escape system (LES), or the rescue rocket. Similar approaches were pursued on the other side of the Iron Curtain. There too, technicians and engineers worked on the development of rescue systems of this type.

The worst possible case was the explosion of a fully fueled rocket on the launchpad or in the first two and a half minutes of the flight, when the rocket and space vehicle were still flying in the dense lower layers of the atmosphere. The use of ejection seats similar to those used in combat aircraft was originally considered. This quickly proved impractical, however, since they were incapable of carrying the astronauts far enough away from the exploding booster rocket. The astronauts would have come down too close to the scene of the accident and would possibly have been struck by falling wreckage or burned in the fireball. The functionality of the ejection seats was also guaranteed only at relatively low speeds. There were no reliable findings as to how they would perform at supersonic and even-higher speeds. And as a rocket broke the sound barrier soon after launch, the point at which ejection seats lost their reliability was quickly passed.

Theoretical considerations led to the conclusion that the most effective way to ensure rescue was to propel the entire spacecraft with its occupants out of the area of the explosion by means of special solid-fuel rockets and then land them by parachute at a safe distance.

Because experience had shown that the probability of the disastrous destruction of the booster rocket decreased with increased flight time, it could be concluded that after a certain time the rescue system would no longer be needed. It was therefore most effective at some point to get rid of the now-excess ballast and jettison it. By then the booster rocket and space capsule were flying in less dense layers of the atmosphere. Should a dangerous situation develop during this phase of flight, the spacecraft could quite simply separate from the booster rocket and, as during a normal landing, return by parachute.

The first Soviet spacecraft were designed without a rescue system. This would finally change with the Soyuz spacecraft with its crew of three. The engineers analyzed a series of different emergency situations with inclusion of a rescue rocket.

To ensure the smooth functioning of the rescue system, the booster rocket with the same name used to launch the Soyuz spacecraft was equipped with an automatic system.

It was capable of detecting all possible malfunctions by the rocket and, with no input from the cosmonauts or ground control, responding automatically by initiating the rescue process.

To further improve safety, a second possibility was envisaged; namely, of activating the rescue system from the launch control center by means of radioed command. Surprisingly, the crew itself was not given the option of activating the rescue rocket by itself.

The system, an improved version of which is still in use today, is based on the idea of propelling the entire space capsule out of the dan-

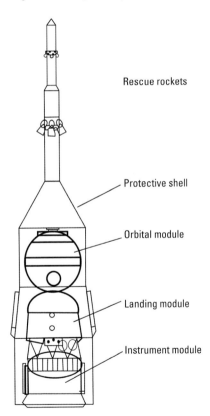

Rescue rockets

Protective shell

Orbital module

Landing module

Instrument module

The three components of the Soyuz capsule in their protective shell, with the rescue rockets at the tip. *Woydt*

During a test, the rescue rockets pull the Soyuz capsule up and away from the launchpad. *RKK Energia*

Rescue from the launchpad: the rescue rocket ignites and pulls the orbital and landing modules away from the rocket. Lattice-type speed brakes slow and stabilize the flight. Finally the landing module is separated and comes down by parachute in a controlled descent.

ger zone with the aid of a rocket propulsion system. The necessary rocket engines should be simple, reliable, and immediately ready to operate in an emergency. They must be able to produce very high thrust for a brief time. The choice was therefore an easy one: only solid-fuel rockets met these requirements.

The rescue system is as simple as it is effective. It is based on several solid-fuel rockets that, mechanically, are part of the aerodynamic shell that surrounds the Soyuz spacecraft during launch. In the event of danger, the thrusters are ignited manually or automatically, and the booster rocket and the spacecraft separate. In a matter of seconds the rescue tower pulls the Soyuz capsule's orbital and landing module away from the rocket. The instrument module stays behind on the booster rocket.

Following the test ignition of a rescue rocket, the landing module has separated and comes down by parachute. *RKK Energia*

Another solid-fuel thruster pointing sideways moved the entire spacecraft laterally to create distance between the spacecraft and the launchpad. After the solid-fuel rockets have burned out and the apex of the parabolic-shaped flight path has been passed, the protective shell and rescue rockets are jettisoned. The landing and orbital modules separate, and the latter descends to earth beneath a parachute at a safe distance from the launchpad.

The rescue system has so far had to prove its capabilities only once, during the launch of the Soyuz T-10 mission, but on that occasion it was vital.

The launch was scheduled for September 26, 1983. In command was thirty-six-year-old air force pilot Vladimir Titov from Siberia. With him was Gennadi Strekalov. The forty-two-year-old civilian and engineer had been to space five months earlier, when he flew with Titov and cosmonaut Alexander Serebrov on Soyuz T-8. They had flown to the Salyut 7 space station, but after several docking attempts failed, the mission was aborted after two days.

Another challenging mission now awaited the two cosmonauts. They were once again to proceed to the space station and try for a second time to dock there. Once on the station, the two were to assist cosmonauts Vladimir Lyakhov and Alexander Alexandrov, who had been on the station for three months, carry out vital repairs. Their tasks included two complicated extravehicular activities to install two additional solar panels on Salyut 7, whose energy consumption had risen sharply over time due to a steady increase in the number of energy-consuming experimental installations.

Following standard Soviet practice, the booster rockets had been mounted while in the horizontal position at the Baikonur Cosmodrome and joined to the Soyuz capsule. Two days before the launch date, the entire assembly was moved on rails to the launchpad and only there was raised into a vertical position. There the rocket and capsule underwent final checks. Nothing unusual came to light, and the countdown began.

The two cosmonauts boarded their spacecraft at 2137 Baikonur time, two hours before launch. During the evening hours the temperature had fallen to 50 degrees Fahrenheit. The wind picked up, with gusts exceeding 25 miles per hour.

The last phase of the countdown was fully automatic; the cosmonauts and ground control limited themselves to monitoring the process. Illuminated by searchlights, the booster rocket stood motionless in the black Kazakh night.

Then, ninety seconds before the end of the countdown, there was suddenly chaos. A fuel valve in the line from the launchpad's supply installation to the booster rocket failed and did not close as anticipated. Nitrogen gas, one of the fuel components, streamed uncontrolled into one of the rocket engine's turbopumps, which began to turn. Its revolutions quickly rose to far in excess of the load limit, and the pump blew apart. The supply lines were destroyed, and fuel streamed out and poured over the lower part of the booster rocket and the launchpad. Then a spark ignited the highly explosive mixture. Within a few seconds, everything was submerged in a sea of flames.

Inside their space capsule, the only indication Titov and Strekalov had at first that something was wrong was unusual vibrations and noises. The protective shield covered the win-

The crew of the Soyuz T-10-1: commander Vladimir Titov (left) and flight engineer Gennadi Strekalov.

A booster rocket with Soyuz capsule is transported to the launch site. Above the spacecraft can be seen the rescue rocket; on the sides just beneath the tip are the four folding stabilizing fins.

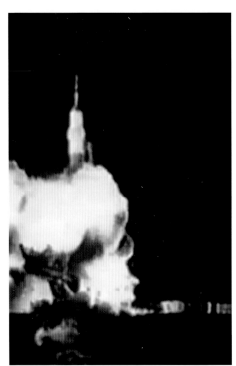
Flames shroud the first stage of the Soyuz booster rocket.

dows of their space capsule, preventing the two men from seeing what was going on around them.

In a bunker over a mile away, the staff of mission control could only watch in horror as the sea of flames grew ever larger. Actually the automatic safety system had long since recognized the looming disaster and should have ignited the rescue rockets, but the flames had destroyed the necessary electrical connections.

The crew's last chance was in the hands of launch control, which would have to manually ignite the rescue rockets by transmitting the ignition command by radio. Two persons in two separate rooms each had to push a button within five seconds. This hurdle had been created as a precaution against the unintentional firing of the rescue rockets.

It was more than ten seconds until the radio command was finally sent. For the first time, the rescue rockets would be fired in a real emergency and not a test situation. With a combined force of 176,475 pounds of thrust, the solid-fuel engines of the rescue rocket powered the orbital and landing module upward. Just seconds later, the booster rocket tipped to the side and exploded in a gigantic fireball.

Within five seconds the Soyuz capsule exceeded the speed of sound, and the two cosmonauts were briefly exposed to a force greater than 14 G. The landing and orbital modules separated at a height of 3,000 feet. The emergency parachute was ejected and the landing module floated to the ground; Titov and Strekalov came down 2.5 miles from the launch site at 2343 local time. The force of impact was significantly greater than that of a normal landing.

Rescue personnel rushed toward the landing site and reached the space capsule thirty minutes later. They found the cosmonauts shaken up, with various blue patches of skin and bruises, but otherwise they were largely uninjured. After the two men had been freed from their capsule, they first asked for cigarettes and a glass of vodka.

The raging fire at the launch site was visible to the naked eye even from that distance. According to witnesses, the remains of the booster rocket burned for more than twenty hours.

Years later, while being interviewed by the American History Channel, Vladimir Titov recalled the dramatic moments: "We had no time to be afraid. The first thing we did when we could think clearly again was turn off the cockpit voice recorder: we cursed!"

At a NASA ceremony celebrating the fortieth anniversary of manned spaceflight, the two cosmonauts expressly thanked Max Faget. His idea had saved their lives.

The Soyuz T-10 mission was finally carried out on February 8, 1984, this time with a crew of three: cosmonauts Leonid Kisim, Vladimir Solovyov, and Oleg Aktov. On April 3 of the following year, Gennadi Strekalov flew to Salyut 7 as flight engineer of the Soyuz T-11. Vladimir Titov was aboard the Soyuz TM-4 when it was launched to the space station on December 21, 1987. Both subsequently continued their careers as space travelers by flying to the Mir space station aboard the American space shuttle.

High-ranking representatives of the Soviet army follow events at the launch site.

A "MAJOR MALFUNCTION": *CHALLENGER* STS-51 L

Wednesday, January 28, 1986, 0630 Florida local time. It was bitterly cold that morning at the Kennedy Space Center in Cape Canaveral, Florida. The sky had cleared during the night, and the temperature had fallen below freezing. An insignificant problem while filling the space shuttle *Challenger's* outer tanks had been addressed quickly. It had nevertheless forced a one-hour postponement of the launch until 1038 local time. A team of technicians had inspected the shuttle and the launch-pad during the night at 0130. Despite the heavy ice buildup and although the temperature was 10 degrees Fahrenheit, it was decided to continue the countdown.

Everything was ready for mission STS-51 L, the twenty-fifth launch by a space shuttle and the tenth by the *Challenger*. In its cargo bay was the TDRS-B, a NASA communications satellite, waiting to be transported into space. The Spartan scientific payload was to be deployed by the crew and independently observe Halley's comet, whose closest approach to the sun was expected on February 5, 1986.

Forty-six-year-old Air Force pilot Francis "Dick" Scobee from Washington State had been chosen to command STS-51 L. Scobee was married and had two children. He had completed his first spaceflight during mission STS-41 C in 1984. The *Challenger's* pilot was navy pilot and space rookie Michael "Mike" Smith from North Carolina. He was forty years old, married, and the father of three children.

The first of the *Challenger* crew's three mission specialists was thirty-six-year-old Judith "Judy" Resnick from Ohio. She had been to space

The crew of the *Challenger* for the STS-51-L mission. Rear row from the left: Ellison Onizuka, Christa McAuliffe, Greg Jarvis, and Judith Resnick. Front row: Michael Smith, Dick Scobee, and Ron McNair.

during a flight by the space shuttle *Discovery*, becoming the second American woman to go into space. Resnick was divorced with no children. The second mission specialist was the Hawaiian Ellison Onizuka, for whom the STS-51 L mission was to have been his second spaceflight. The forty-year-old married father of two had flown on the shuttle before, during the military shuttle mission STS-51 C in January 1986. Ronald "Ron" McNair from South Carolina, at thirty-five the youngest member of the crew, completed the list of mission specialists. The coming flight by the *Challenger* would be the first mission for the married father of two children.

The crew also included two payload specialists. One was forty-one-year-old Gregory "Greg" Jarvis from the state of Michigan. For Jarvis, who was married, STS-51 L would be his first space mission. The second payload specialist, Massachusetts-born Christa McAuliffe, had attracted much attention during the run-up to the mission. McAuliffe was a teacher at a high school in New Hampshire. Together with 11,000 others, she applied for the Teacher in Space program. During the spaceflight she was to give two half-hour lessons broadcast live on television, one called "The Ultimate Field Trip" and the other called "Where We've Been, Where We're Going, Why." Christa McAuliffe was married and had two children.

NASA was constantly under pressure. In the twenty-fifth year of the space program the high expectations of the shuttle program were slowly but surely giving way to reality. The original plan had envisaged the small fleet of orbiters making flights at almost weekly intervals, delivering payloads to space at unbeatably low cost. NASA hoped to secure a market advantage in satellite launches over ambitious competitors from Europe, Russia, and other ris-

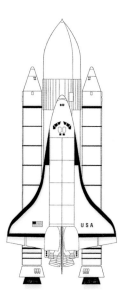

Space shuttle in launch configuration. *NASA*

ing spacefaring nations.

It had already been shown that the shuttles, in their day the most-complex machines ever built by man, were subject to technical problems. Maintenance and repairs after a mission had proved more time consuming and costly than originally hoped. As a result, the intervals between flights lengthened dramatically, As well, ever more frequently, flights had to be postponed for technical reasons, often at the last minute.

After initial successes such as spectacular satellite repairs in orbit, the transition to the strived-for routine operation did not take place. It seemed that the hope that the shuttle program could ever meet costs or even make money might soon have to be abandoned.

Preparations for mission STS-51 L, originally scheduled for January 22, 1986, began under these unfavorable conditions. The first adjustment of this date had to be made when the preceding flight by the Columbia had to be delayed. On December 19, 1985, its onboard computer stopped the countdown fourteen seconds prior to ignition due to hydraulic problems. NASA was forced to postpone the launch of *Columbia*, initially to January 4 and then to January 6. This also affected the timetable for the *Challenger*. The launch was initially rescheduled for January 23 the following year and finally by another day to the 24th.

The problems with *Columbia* were far from over, however. On January 6 a defective fuel valve prevented the launch, and the next day the weather was unfavorable. On January 9 a defective temperature sensor was found in the fuel system, and the next day it poured rain. Not until January 12 did the orbiter succeeded in launching.

After all these delays, the launch of the *Challenger* was set for January 25, 1986, but then it was postponed by another day. This date also soon became uncertain, however. Weather conditions in Florida deteriorated rapidly. A cold front with unusually low temperatures for that part of the country had arrived, and a thick layer of clouds was approaching the Kennedy Space Center.

The day before the launch, the weather forecast for the next day was anything but ideal. For safety's sake the launch was postponed by another twenty-four hours until the following Monday, January 27.

On the night of Sunday–Monday the final phase of the countdown finally began, with the filling of the orbiter's external tank with liquid hydrogen and liquid oxygen. Those involved in the preparations watched what was happening with skepticism. Ice was rapidly building up on the orbiter and the launch facility.

The crew of the *Challenger* had been wakened shortly after five on the morning of the launch. At about the same time, renewed checks of the shuttle and the launch installation were carried out. Nothing had changed since the previous inspection four hours earlier. The formation of ice had not decreased—on the contrary. The layer of ice on the struts and lines had become even thicker. Management was confident, however, that the morning sun would melt most of the ice.

After a crew breakfast, the men and women of the crew put on their sky-blue flight suits and to the applause of NASA staff and journalists made their way to the launchpad. There they took the elevator up to the White Room, the preparation room from which all astronauts boarded their spacecraft. The crew took their seats in the crew compartment, with Commander Dick Scobee in the left front seat and Pilot Mike Smith in the seat to his right. Sitting behind the two was Judy Resnick. As flight engineer, it was her task to help the commander and pilot monitor the instruments during the

Ice buildup on a railing of the launch facility on the morning of January 28, 1986. *NASA*

Assembled solid-fuel booster on a special transporter. *NASA*

launch phase and help them work their way through the extensive checklists. Beside her sat Ellison Onizuka. Christa McAuliffe, Greg Jarvis, and Ron McNair had taken their places on the lower deck.

After ground personnel helped the crew strap in, they connected their helmets to the shuttle's oxygen system, and each also connected to a separate personal egress air pack (PEAP) for emergencies. These suitcase-size oxygen containers were supposed to help the astronauts escape the danger zone in the event of an emergency abort and the possible release of poison gases.

To give the inspection team another opportunity to have a close look at the ice formation, the launch was again delayed, this time until 1138 local time. In a statement, NASA public affairs officer Hugh Harris said: "One of the considerations concerned the icicles, some of which were several feet long and during launch might possibly break off and damage the orbiter's heat shield."

This was the first outward sign that the unusual formation of ice had garnered the attention of those responsible.

The final phase of the countdown began at 1129, nine minutes before launch. The White Room was swung away from the *Challenger's* hatch, and Pilot Smith engaged the shuttle's power generator. After a test of the engine nozzle control and the elevators and ailerons, thirty-one seconds prior to ignition the computer

Challenger lifts off on mission STS-51-L. No one suspected that disaster was just seconds away. *NASA*

took control of events, and Harris announced: "We have a GO for automatic launch."

The self-destruct mechanisms of the solid-fuel rockets, which in case of an accident could destroy the so-called booster, were armed eleven seconds prior to liftoff. In contrast to liquid-fueled rockets, solid-fuel rockets have the dangerous property that once ignited they cannot be turned off.

Just one second later the computer in the control center gave the ultimate GO for launch. Harris counted down the last seconds: "10, 9, 8, 7, 6 . . ." The shuttle's three main engines thundered to life at T = 0, 6.6 seconds prior to liftoff. The computer confirmed the proper functioning of all three engines, and while Harris counted down the last remaining seconds, "3, 2, 1," at zero the solid-fuel rockets ignited. Just 250 milliseconds later the *Challenger's* computer registered the first vertical movement.

Judith Resnick could be heard on the shuttle astronauts' internal intercom: "Yahoo!" Smith answered: "Here we go!"

Just seven seconds later, *Challenger* had left the dangerous proximity of the launch tower behind it. The control center in Houston took over from launch control at Kennedy Space Center. NASA public affairs officer Harris declared: "Launch of the twenty-fifth space shuttle mission and the shuttle has cleared the tower."

The output of the main engine rose as planned to 104 percent, and the roll maneuver was initiated. The shuttle rolled about its longitudinal axis, after which it, so to speak, hung beneath the huge external fuel tank as it rose into the sky. This maneuver reduced aerodynamic forces on the orbiter and set it on course over the Atlantic as it climbed.

Eight seconds after liftoff, Commander Scobee reported the beginning of the maneuver: "Houston, *Challenger*—Roll program." Astronaut Richard "Dick" Corvey, who was acting as CAP-COM (CAPsule COMmunicator) in the control center, acknowledged, "Roger, roll *Challenger*."

Eleven seconds later Pilot Smith informed his astronaut colleagues over the shuttle's intercom: "Looks like we've got a lotta wind here today." Scobee replied with a brief "Yeah." NASA later confirmed that the crosswind was stronger than any encountered by previous shuttles during launch.

The *Challenger's* computer began reducing thrust of the three main engines to 65 percent as per the flight plan. Just twenty-eight seconds after liftoff the *Challenger* passed the 10,000-foot mark and was at half the speed of sound. Another twelve seconds later it passed an altitude of 19,000 feet and exceeded the speed of sound. At T = 48 seconds the engines had reached 65 percent thrust. Just three seconds later they received the command to return to full power. Sixty seconds after liftoff, Smith advised the astronauts over the intercom: "34,000 feet, reaching 1.5 [times the speed of sound]."

At T = 68 seconds the instruments in the control center in Houston and in *Challenger's* cockpit showed that the engines were again operating at a nominal thrust of 104 percent. CAPCOM Covey transmitted, "*Challenger*, you are go at throttle up." Seconds later, Com-

Seconds after the explosion. A solid-fuel rocket shoots from the cloud of wreckage left by the *Challenger. NASA*

mander Scobee acknowledged, "Roger, go at throttle up." In Houston, NASA spokesman Steve Nesbitt, who had taken over commenting on the launch for the press and public from his colleagues at the Kennedy Space Center, announced this over NASA's own television and news network.

Then suddenly, things began happening quickly. Pilot Mike Smith was heard to say over the intercom: "Uh oh!" Seventy-three seconds had passed since launch. At that time the shuttle was at an altitude of approximately 46,000 feet at almost twice the speed of sound. Thousands of dismayed observers, including many family members of the crew, could only watch as from one moment to another the *Challenger* was shrouded in a sea of flames and a huge cloud of white smoke, from which the two solid-fuel rockets, each trailing a jet of flame, shot out, zigzagging wildly.

Obviously unaware of what had happened, in the control center NASA spokesman Nesbitt announced: "One minute fifteen seconds. Velocity 2,900 feet per second [1,977 miles per hour]. Altitude 9 nautical miles. Downrange distance 7 nautical miles."

At the Kennedy Space Center, witnesses to the event saw large and small pieces of wreckage emerge from the white cloud left by the explosion, and, leaving white condensation trails behind them, plunge toward the surface of the Atlantic off the coast of Florida.

At T = 86 seconds, thirteen seconds since what had obviously been an explosion, Flight Director Jay Greene reacted to the pictures on his monitor. He turned to the controller responsible for flight tracking (FIDO = Flight Dynamics Officer), asking him for information about the shuttle's flight path: "Flight, FIDO, filters got discreting sources. We're go." Then, "Flight,

Flight Director Jay Greene (behind) and Alan Briscoe watch uncomprehendingly the events in the sky over Florida on a television monitor. *NASA*

GC, we've had negative contact, loss of downlink." Greene said to his people: "OK. All operators, watch your data carefully."

Thirty-seven seconds after the explosion, the flight controller responsible for safety on the ground transmitted the order for the destruction of the solid-fuel rockets, which were still racing about in the air, in order to prevent them from possibly coming down in a populated area.

One minute and fifty-six seconds after the liftoff of the *Challenger* (T = 1 minute 56 seconds), spokesman Nesbitt informed the public: "The flight controllers are watching the situation very carefully." And then the understatement of the century: "Obviously a major malfunction." Years later in an interview in the magazine *Popular Mechanics*, Nesbitt declared that he had no reliable information at that time. Although he suspected that the shuttle had exploded and that the crew could not have survived, out of journalistic professionalism he did not want to make such a definitive statement until it was confirmed by official sources.

Two minutes and twenty-five seconds after the launch of the *Challenger* the FIDO reported to

The committee to investigate the *Challenger* disaster arriving at the Kennedy Space Center. In the center is William Rogers. *NASA*

Greene: "Flight director, FIDO here." Greene told him to go ahead, and the FIDO confirmed what everyone already knew: "The ground safety controller reports that the vehicle has exploded."

There was a lengthy silence, then the helpless question from the flight director: "Understood, FIDO. Can we get any information from the rescue teams?" He received no answer.

At T + 2 minutes 45 seconds, Greene told his staff: "Ground control, emergency procedures are in force." This required the flight controllers to seal off the control room. The doors were locked and incoming and outgoing phone calls were prohibited. At the same time, the staff was directed to secure all available information that might possibly help in the investigation into the incident.

Steve Nesbitt turned again to the public: "We have a report from the controller responsible for monitoring the flight path that the vehicle has exploded. The flight director confirms this. We and the rescue forces are checking to see what can be done at this time."

Meanwhile, television images clearly showed pieces of wreckage striking the surface of the Atlantic Ocean. Rescue teams rushed to the scene in boats, helicopters, and aircraft, but all of those who were familiar with the subject already knew with certainty that there could not have been any survivors.

The first survey made by the flight controllers showed nothing out of the ordinary in the telemetry data received from the shuttle and its propulsion units until radio communications were lost. Although the rescue of possible survivors had the highest priority, attention was also already being turned to finding the cause of the accident. The catastrophe had played out before the eyes of the world public—the biggest disaster in manned spaceflight to date had taken place live on television. Not only the American nation but the whole world was in shock. That evening American president Ronald Reagan turned to the nation with an emotional speech that ended with the words: "Nothing ends here; our hopes and our journeys continue."

First, however, the circumstances of the accident had to be explained. President Reagan appointed an investigating committee, frequently called the Rogers Committee after its chairman, the former politician and diplomat William P. Rogers. The committee had fourteen members, including former astronaut Neil Armstrong, the first man in the moon, and Sally Ride, the first American woman in space. It also included flying legend Chuck Yeager, who had made history as the first man to fly faster than the speed of sound, and Nobel Prize–winning physicist Richard Feynman. The committee began its work immediately. What it found was plainly and simply unbelievable.

First they reviewed the extensive data, including sound and video recordings, of the accident in order to understand the course of events in detail. The first indication that something was out of the ordinary on this launch was identified at 0.678 seconds after the ignition command for the solid-fuel rockets, in images of the launchpad. A black cloud of smoke could be seen. It seemed to originate from the lower part of the right solid-fuel booster, precisely at one of the places where this multipiece external shell had been jointed. Also, the location was near the rear of the two struts that attached the booster to the liquid-hydrogen-filled external tank of the shuttle. The black cloud was seen for approximately 2.5 seconds, then it disappeared from the images.

A dark cloud of smoke is visible on the inner side of the right solid-fuel rocket seconds after ignition—hot gases are escaping from the booster. *NASA*

The collected data showed that at T = 37 seconds the shuttle's computer had reacted to the prevailing wind conditions and kept the vehicle on course by counter-steering. Forty-five seconds after liftoff, a flash was seen on the outer side, three seconds later on the trailing edge of the orbiter's right wing. Fractions of a second later, a bright red flame appeared beneath the right wing in close-ups taken by a telecamera. The flames quickly merged with the nozzle jet from the solid-fuel booster on that side. This phenomenon had been observed during other shuttle launches and had never caused any damage. It seemed to have no causal connection to the accident.

After just a minute of flight, camera images showed the first signs of an unusual appearance of flames on the side of the right solid-fuel rocket facing the shuttle. The flames became increasingly visible in the following seconds, and a regular jet of flame directed away from the shuttle formed. Simultaneously, telemetry data showed a rapid pressure drop inside the

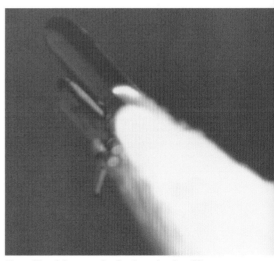

About fifty-eight seconds after launch, a jet of flame emerges from the rear end of the solid-fuel booster. *NASA*

booster, a clear indication that the hull must have developed a leak.

At T = 60.248 seconds, flames first became visible that were obviously directed at the shell of the shuttle's external fuel tank. They struck it in the area of the rear connection between the tank and the right solid-fuel booster. The flames

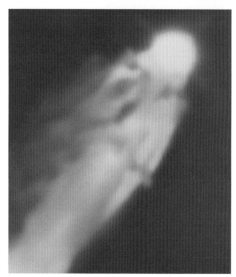

The moment of the catastrophe: liquid oxygen and hydrogen from the external tank have escaped and exploded. *NASA*

the flames were being fed by an additional source. The loss of pressure caused the thrust produced by the right solid-fuel booster to drop. The asymmetrical thrust profile caused the shuttle to begin deviating from its course. The onboard computer tried to compensate by pivoting the thrust nozzle of the shuttle's main engine.

At T = 65.164 seconds the telemetry for the first time began indicating unusual movements by the vehicle. Once again the onboard computer tried to compensate.

Now the number and intensity of the fires rose in a very short time. From T = 66.764 seconds, telemetry showed a rapid drop in hydrogen pressure in the shuttle's external tank, a clear sign that there was also a leak in this tank. Seconds later, pressure in the 17.7-inch-diameter feed line for liquid oxygen to the main engines began to drop.

After 72.204 seconds of flight time the data transmission showed divergent movements of the engine nozzles of both solid-fuel engines. A moment later the right of the two boosters moved, which could be explained only by the destruction of the rear support strut between the booster and the external tank. At that moment the television pictures showed a large, rapidly growing fireball in the forward third of the shuttle. Obviously, the tip of the booster had dented and then torn open the external fuel tank, resulting in leakage. Hectic control movements by the solid-fuel engine nozzles and all three main engines followed. Just seconds later, telemetry showed a sideways movement by the orbiter; 72.525 seconds had passed since launch. It is suspected that this movement was the first unusual observation of which the astronauts were conscious.

The pressure of the liquid hydrogen and oxygen fell rapidly, and the powerful control

remained visible in the seconds that followed.

Two seconds later, the *Challenger's* onboard computer reacted a second time to the crosswind. It initially moved the nozzle cone of the left booster, and two-tenths of a second later it moved the elevator on the right wing.

At T = 64.660 seconds the jet of flame in the direction of the external tank suddenly changed its intensity. A consistently bright glow was recognizable; obviously, from that moment on,

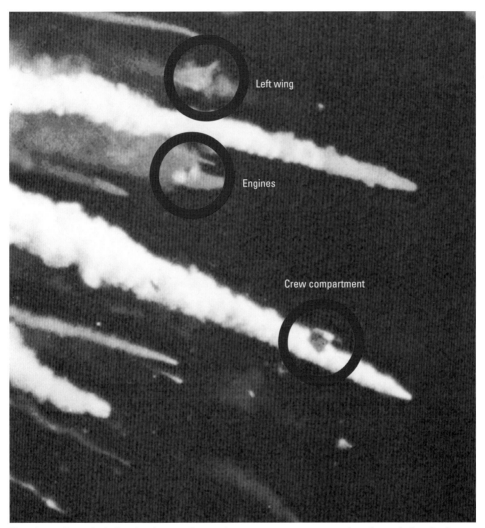

Left wing

Engines

Crew compartment

Three large pieces, including the almost undamaged crew compartment, emerge from the explosion cloud.

movements continued. The orbiter moved sharply to the left. The thrust difference between the right and left solid-fuel boosters reached a dramatic scale. A bright cloud at the lower end of the external tank suggested a correspondingly large leak in the hydrogen tank. Shortly thereafter the liquid oxygen tank also began losing fuel.

After 72.213 seconds of flight, there was an explosion near the forward part of the solid-fuel rockets, which increased in intensity very rapidly. Telemetry data confirmed that main engine fuel pumps had reached their load limits. Then everything happened extremely quickly: the last useful telemetry data was received 73.631 seconds after launch. Just 0.499 sec-

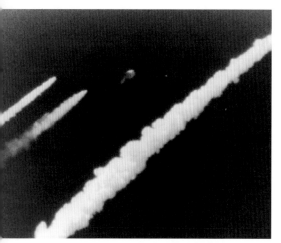

This enlargement shows the *Challenger's* crew compartment, which seems largely intact. It continued climbing for several miles before plunging into the ocean on a parabolic trajectory. *NASA*

onds later the transmission stopped. After another half second, film images showed the forward part of the orbiter with the crew compartment shrouded by a bright red fireball. Obviously, the fuel from the orbiter's maneuvering thrusters had escaped and exploded. The cloud expanded rapidly to cover the entire orbiter.

The photos show that while the shuttle's external tank had exploded, the orbiter itself had not. The shuttle had survived the explosion more or less intact, although it broke up soon afterward due to sudden powerful aerodynamic forces. Large pieces could be identified in enlargements, including the unit with the three still-functioning main engines, one wing, and a large part of the fuselage.

The crew compartment was also clearly identifiable. For the moment it was intact, and in the following twenty-five seconds its own momentum carried it from the position of the explosion at an altitude of 47,900 feet to almost 65,600 feet. It then followed a parabolic

flight path, in free fall, toward the surface of the Atlantic Ocean. The crew compartment of the *Challenger* crashed into the waves at more than 180 miles per hour three minutes and fifty-eight seconds later.

Deformation of the astronaut seats, which were recovered later, showed that the impact must have taken place nose first, facing to the right. The forces that resulted from the impact exceeded 200 G. This absolutely would have been fatal to any crew members who had survived thus far. Despite the heavy impact, the cabin remained largely intact. It sank with the bodies of the astronauts to the seafloor, 100 feet down.

Aircraft and helicopters, which were already on their way to the crash site, at first had to be held back out of fear that they might be struck by falling debris. There was no hope of survivors in any case. In the end, all that remained was to collect as much wreckage as possible from the water for analysis in an extremely costly recovery effort. In particular, the physical remains of the astronauts had to be found.

In mid-March, one and a half months after the accident, recovery crews made a shocking discovery. They succeeded in recovering four of the seven PEAPs, the astronauts' emergency oxygen containers, from the sea. Their examination gave a possible clue to the fate of the astronauts. It had previously been assumed that the crew had been killed in the explosion or moments later had been suffocated or at least rendered unconscious seconds later by lack of oxygen. Now the accident investigators found that three of the four PEAPs had been activated. These devices could only be activated manually, however. This led to the conclusion that at least some of the astronauts were fully conscious after the explosion and must have been capable of acting. They must also have realized what had happened and what was about to happen.

On April 18, NASA was finally able to announce that the crew cabin with the physical remains of all seven crew members had been found and recovered under conditions of strict secrecy. The remains were then forensically examined.

During a press conference held on July 28, NASA made public a report by Joseph P. Kerwin, a doctor and former astronaut. He stated his opinion that the cause of death of the crew would probably never be determined with certainty. Nevertheless, he stated that when the orbiter broke up, it experienced temporary forces of 12 to 20 G, which had probably not been strong enough to cause the immediate deaths of the astronauts. Further, he put forward the thesis that the crew had probably lost consciousness due to lack of oxygen, although this could not be proved with certainty. Impact with the water after a fall of two minutes and forty-five seconds from an altitude of almost 12 miles could not have been survived. In conclusion, Kerwin confirmed that the examining pathologists had been unable to determine if lack of oxygen had caused the deaths of the seven crew members. No other cause of death could be established with certainty.

Details of the autopsy reports, such as photos of the recovered crew compartment, were not made public out of respect for the dead astronauts and their families.

By mid-August the investigators were certain that they had found the cause of the accident. The search for further wreckage from *Challenger* was halted. It had lasted seven months, covered an area of 13,500 square miles, and cost many millions of dollars. As many as 6,000 people had taken part in the search effort; 45 percent of the orbiter and 50 percent of the external tank and the two solid-fuel rockets had been recovered from the sea, as well as an ever-larger portion of the payload from the shuttle bay.

In searching for the cause, three questions in particular concerned the Rogers Committee: What had been different about the launch of the *Challenger* compared to the twenty-four successful shuttle launches that preceded it? What pointed to the right solid-fuel booster rocket as the cause of the accident? What was the precise chain of events that ultimately led to the explosion?

The unusual black smoke cloud at the bottom end of the right booster after its ignition was a first starting point for the investigators. After ignition of the solid-fuel booster, which during the investigation was reassembled from several segments, under the pressure from the burning fuel its various parts moved against each other inside the booster. The sections

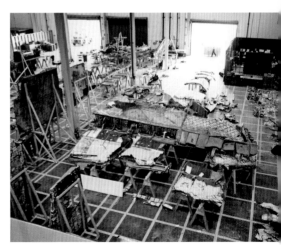

Pieces of wreckage from *Challenger* in a hangar at the Kennedy Space Center. *NASA*

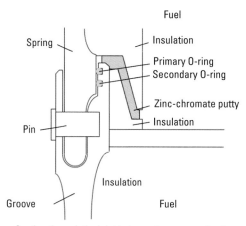

Section through the joint between two segments of a solid-fuel booster. *NASA/Woydt*

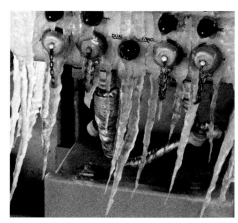

Accumulated ice on armatures on the launchpad. *NASA*

were joined by nuts and bolts, and the joint was sealed by rubber rings.

Because of the tremendous forces during launch, these rubber rings were squeezed, stretched, decompressed again, and so forth, in ways that were difficult to anticipate. As temperatures dropped, however, the ability of the rubber to return to its original shape was drastically reduced. The well-below-freezing temperatures during the night and the air temperature at the time of launch, just 35.6 degrees Fahrenheit, had caused the rubber rings to become brittle and stiff. This fact was impressively demonstrated by Richard Feynman, a member of the investigating committee, in a hearing on live television. He first demonstrated the elasticity of a piece of the rubber used to join the booster segments. Then he placed it in a glass of ice water for a few seconds. The result of this small but impressive experiment was that the piece of rubber became so hard that it lost is ability to return to its original shape.

This is exactly what had happened at the moment the solid-fuel boosters were ignited. Several O-rings in the right booster failed to

completely seal the gaps in the interface between two segments. The escaping combustion gases had melted the ring. This could be seen in the black smoke clouds in the film images taken during launch. Just a few seconds later, combustion residue, a sort of solid slag, had closed the hole again. However, when the shuttle was shaken by strong wind gusts as it passed through the atmosphere, the hole in the segment joint was opened again, this time for good, by the powerful vibrations and movement of the spacecraft. Hot combustion gases now had an unhindered path to the outside.

The cause of the *Challenger* accident had been found. The jet of flame caused by the leak had struck the rear part of the orbiter's external tank and the rear support brace between it and the booster. The wall of the hydrogen tank in the external tank had literally been cut open, and the contents of the tank poured out. At the same time, the bracing strut between the external tank and booster burned through. The booster, still held by the upper strut, turned about this pivot point. Its tip bored into the front part of the external tank and also caused a leak there. Oxygen now flowed from this hole, mixing with the hydrogen from the rear part to form a highly explosive gas. The result was a tremendous explosion with the known catastrophic results for the orbiter and its crew.

What came to light about the history leading up to the accident revealed a scandal, a long sequence of sloppiness, cover-ups, and incompetence. The reliability of the solid-fuel rocket design had been questioned in the early 1970s, at an early stage in shuttle development. Experiments had shown that the proposed use of rubber rings by the manufacturer of the booster rockets, Morton Thiokol, could lead to leaks. The pressure of the burning fuel inside the booster caused uncontrollable movements of the booster segments against one another, which could not be completely compensated for by the O-rings. It was already clear at that time: the burning through of the sealing rings combined with the escape of hot combustion gases could not be completely excluded. This would undoubtedly lead to the loss of the shuttle and its crew.

NASA engineers warned their superiors about the danger, especially because of the fact that there was no guarantee that the rings would function at low temperatures. Their information was simply ignored and was not passed on to subcontractor Morton Thiokol. Instead the flawed design was accepted by NASA in 1980.

During the first shuttle launches, there had in fact been dangerous changes in the O-rings. This fact was subsequently added to the highest risk management stage, but once again NASA management was not informed. No one came up with the only correct response: to halt shuttle flights until the problem was solved. Finally, in the mid-1980s Morton Thiokol dealt

with the problem itself. The design of the joining elements between the segments was improved, and Thiokol informed NASA that the problem had been solved. NASA was all too ready to accept this, with the well-known catastrophic consequences.

After the weather forecast for the night and morning of January 28, 1986, predicted such low temperatures, NASA employees remembered the warnings about the O-rings and their behavior at low temperatures and contacted the supplier. Company technicians called for a postponement of the launch until the warmer afternoon hours. The night before the launch there was a spontaneous telephone conference involving managers and engineers from the participating parties, Morton Thiokol and NASA. During the conference call the manufacturer again warned forcefully against the launch in the forecasted low temperatures and requested that the launch be delayed. At the beginning of

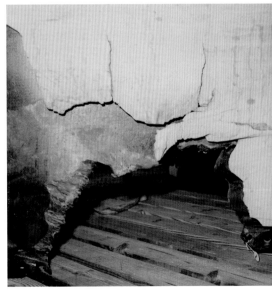

Wreckage from the right solid-fuel booster. The hot gases literally melted a hole in the outer casing. *NASA*

The physical remains of the *Challenger* crew were transported in a US Air Force transport aircraft. *NASA*

the conference call the Thiokol engineers were still supported by their management.

NASA personnel argued against canceling the launch. In their opinion, if one of the two O-rings failed during launch, the second ring would ensure that the system remained secure. This claim had never been investigated in practice, however. In addition, this decision-making process—namely, trust in the second O-ring as a backup system—contradicted NASA guidelines for dealing with a safety risk of the highest level.

Perhaps not least because new contracts were to be concluded in the near future, Thiokol management finally gave in to the pressure and supported the planned launch time. Senior NASA management was not advised of the conference call and its questionable result. Not until much later did the unholy processes come to the light of day during research into the cause of the accident.

The physical remains of the seven astronauts were released to their families for burial on April 29, 1986.

Commander Dick Scobee, pilot of the unlucky mission, and Mission Specialist Judy Resnick were buried at Arlington National Cemetery in Arlington, Virginia. The remains of Ellison Onizuka were laid to rest at the National Cemetery of the Pacific in Honolulu, Hawaii. Ron McNair was buried in his home town of Lake City, South Carolina. Greg Jarvis was cremated and his ashes were spread over the Pacific Ocean. Christa McAuliffe was buried in her hometown of Concord, New Hampshire. Remains that could not be identified were buried together at the Space Shuttle Memorial at Arlington National Cemetery on May 20, 1986.

Most of the recovered wreckage was buried by NASA in two abandoned rocket silos on the grounds of the Kennedy Space Center, and on February 23, 1987, these were sealed with heavy concrete covers. Should it ever be necessary, these covers could be removed by cranes in order to remove the more than 100 tons of wreckage.

All subsequent shuttle flights were post-poned for an indefinite time during the investigation into the *Challenger* disaster. Not until September 29, 1988, did a space shuttle finally take off again for space, after a break of thirty-three months. All went well until almost exactly seventeen years after the explosion of the *Challenger*, when luck abandoned NASA once again.

The remains of the space shuttle *Challenger* are stored in a missile silo at the Kennedy Space Center. *NASA*

Sequence of Events

Time (Cape Kennedy Local Time)	
11:37:53	Ignition of *Challenger's* three main engines.
11:38:00.010	Ignition of the solid-fuel rockets.
11:38:00.260	The shuttle lifts off.
11:38:00.688	A black cloud of smoke is visible at the lower end of the right solid-fuel rocket.
11:38:00.846	In the next two seconds, eight successive smoke clouds are visible.
11:38:04.349	Main engines go to 104% power.
11:38:05.684	Pressure in the right solid-fuel rocket is 117 psi higher than normal.
11:38:07.734	Roll program begins.
11:38:16	Pilot Smith reports strong crosswind.
11:38:19.869	Main engines are at 94% power.
11:38:21.134	Roll program ends.
11:38:35.389	Main engines go to 65% power.
11:38:37.000	*Challenger's* computer corrects for crosswind.
11:38:45.227	A bright light becomes visible at the rear end of the right wing.
11:38:48.128	A second light becomes visible on the front of the right wing.
11:38:48.428	A third light is visible on the rear end of the right wing.
11:38:51.870	Main engines go to 104% power.
11:38:58.798	First indication of a jet of flame are seen on the right solid-fuel rocket.
11:38:59.272	Sharply delineated flames are steadily visible on the right solid-fuel rocket.
11:38:59.763	Flames spreading in the direction of flight.
11:39:00.014	Pressures inside the left and right solid-fuel rockets diverge sharply.
11:39:00.258	First indications are seen that the jet of flame is striking the shuttle's external tank.
11:39:02.494	Brief, powerful control movement of the right elevator.
11:39:04.670	The flames change their appearance.
11:39:04.715	Steady light on the side of the external tank.
11:39:04.947	The computer moves the main engine's nozzle to hold the shuttle on course.

11:39:05.174	The shuttle begins vibrating heavily.
11:39:06.774	Flow rates in the fuel line for liquid hydrogen from the external tank to the main engines begin to fluctuate.
11:39:10	Last radio message from the crew.
11:39:12.214	Major control movements by the nozzles of the three main engines and the two solid-fuel rockets.
11:39:12.574	Hydrogen pressure in the shuttle's external tank drops.
11:39:12.974	Rapid pressure loss in the liquid oxygen fuel line.
11:39:13.020	Last valid telemetry data are received.
11:39:13.054	Rapid pressure loss in the liquid hydrogen fuel line.
11:39:13.134	Bright light on the rear part of the external tank.
11:39:13.147	The first indications of gases in the area of the external tank become visible.
11:39:13.153	Engines react to changes in fuel flow.
11:39:13.172	The tip of the right solid-fuel rocket strikes the external tank.
11:39:13.201	Bright light becomes visible in the area between the external tank and the shuttle. Bright light becomes visible in the area of the forward attachment between the shuttle and the external tank.
11:39:13.223	The light between the external tank and the shuttle grows in intensity.
11:39:13.292	First sign of a white light in the area of the forward attachment between the shuttle and the external tank becomes visible.
11:39:13.337	Sharp increase in the intensity of the white light in area of the forward attachment between the shuttle and the external tank.
11:39:13.387	Pressure variations begin in the area of the control nozzles.
11:39:13.393	Temperatures of all three main engines are in the red zone.
11:39:13.492	The three main engines shut down in succession.
11:39:14.140	Radio communications with shuttle lost.
11:39:14.597	Bright light in the area of the shuttle's nose.
11:39:50.260	Self-destruct order transmitted for right solid-fuel rocket.
11:39:50.262	Self-destruct order transmitted for left solid-fuel rocket.

The Space Shuttle's Solid-Fuel Boosters

From ignition until they burned out, the two solid-fuel rockets (solid rocket boosters or SRBs) delivered more than 80 percent of the space shuttle / tank / booster combination. Each booster produced 3,304,691.50 pounds of thrust, the rest being produced by the

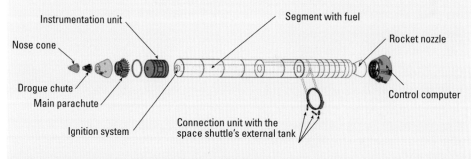

Schematic composition of a solid-fuel booster. *NASA*

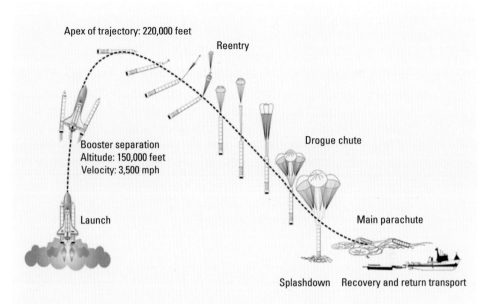

Apex of trajectory: 220,000 feet

Reentry

Booster separation
Altitude: 150,000 feet
Velocity: 3,500 mph

Drogue chute

Launch

Main parachute

Splashdown   Recovery and return transport

Sequence of the launch, jettisoning, and splashdown of the solid-fuel rockets. *NASA/Woydt*

three engines of the orbiter, which were powered by a mixture of liquid hydrogen and liquid oxygen.

Each SRB was 149 feet long and 12 feet in diameter. At launch, each weighed 650 tons, of which 550 tons was fuel. They were manufacturer by Morton Thiokol in its factory in Brigham City, Utah.

A solid-fuel booster consists of a nose section, the rocket engine, and the exhaust nozzle. The rocket engine comprised eleven steel sections: the forward terminating segment, six cylindrical segments, the ring (by which the SRBs were mechanically attached to the shuttle's external fuel tank), two stiffening segments, and the aft terminating segment.

The eleven segments were joined together by using circumferential tang, clevis, and clevis pin fastening, and the joints were sealed by just two 0.27-inch-thick rubber O-rings. Each individual joint was secured by 177 steel pins.

At the factory the eleven segments were assembled into four larger components, into which the fuel was placed. These four components were transported by rail to Cape Kennedy, where they were assembled and joined to the nose and exhaust sections. Then the two solid-fuel boosters were attached mechanically and electrically to the external tank and thus to the orbiter with the aid of two lateral sway braces and a diagonal attachment.

During the launch of the shuttle, its three main engines ignited first. Only when the launch computer signaled after 6.6 seconds that each had reached at least 90 percent of anticipated power were the boosters also ignited. If this was not the case, the shuttle's engines were shut down again. The SRBs had the same disadvantage as all solid-fuel power plants: once ignited, they could not be shut down again; they operated until their

A burned-out solid-fuel booster comes down in the sea. *NASA*

fuel was expended. That meant that they could not be ignited until all other systems were in the green.

The SRBs burned for 125 seconds before their fuel was consumed. At an altitude of 28 miles, two flaps on the side of each were blown out, equalizing the thrust from both boosters. This was necessary for separation, since asymmetric thrust distribution would have pushed the orbiter off course and inevitably destroyed it. Not until this thrust equalization had been completed were the sway braces connecting the boosters to the external tank separated by explosive charges. Finally, small thrusters forced the solid-fuel rockets away from the shuttle, after which they fell behind. Their momentum carried them upward for another seventy-five sec-

onds, achieving additional altitude of about 12 miles, until at about 40 miles they passed the apex of their trajectory and then fell back to earth on a ballistic parabolic trajectory.

Each booster released a drogue chute, which stabilized its flight path and at a predetermined height pulled three large parachutes from their container. In the Atlantic, two recovery vessels waited to pick up the STBs after they splashed down. They were then taken ashore, completely disassembled, checked for damage, and, if appropriate, repaired. Then, after the segments had been filled with fuel, the boosters were ready for the next mission.

FATAL RETURN: *COLUMBIA* STS-107

At 0810 Houston local time on the morning of Friday, February 1, 2003, in Mission Control, the CAPCOM gave the GO for the space shuttle *Columbia* to return to earth.

This was *Columbia's* sixteenth day in space. Mission STS-107 had begun on January 16 with what appeared to be a picture-book launch. During the mission, scientific research had been carried out, principally in the fields of physics, materials science, earth observation, biology, chemistry, and medicine, with the focus on the crew of seven—five men and two women.

The crew was led by forty-five-year-old Air Force officer Rich Husband of Texas. The married father of two children had been to space in 1999, when the crew of the STS-96 mission delivered supplied to the International Space Station. The shuttle's pilot was forty-one-year-old William "Willie" McCool from California. For the married McCool, this flight was his first space mission.

The crew included four mission specialists. One was David Brown, an unmarried forty-six-year-old space rookie from Virginia. Born in India, Kalpana "K. C." Chawla was married with no children. She had also flown on the space shuttle before, in 1997. The third mission specialist was Air Force pilot Michael Anderson from New York State. At the time of the mission he was forty-three years old. Married and the father of two children, this was his second time in space. His previous mission had been on the orbiter *Endeavour* in January 1986. The fourth specialist was Laurel Clark, a doctor born in Iowa. Married with one son, she was forty-one and making her first

Launch of the space shuttle *Columbia* on January 16, 2003, on mission STS-107. *NASA*

spaceflight. The seventh and last member of the crew held a special status: Ilan Ramon was a forty-eight-year-old Israeli. Born in Jerusalem, he was a former combat pilot, married, and the father of four children. On the *Endeavour* his role was payload specialist. STS-107 was his first space mission.

The crew of the STS-107 mission. Top row from left: David Brown, Willie McCool, Mike Anderson. Bottom row: Kalpana Chawla, Rick Husband, Laurel Clark, and Ilan Ramon. *NASA*

Commander Husband and his pilot, Mc-Cool, had readied everything for reentry into the earth's atmosphere. Weather conditions at the Kennedy Space Center in Florida, where the orbiter was to land, were acceptable. Everything seemed in order. The shuttle's two maneuvering thrusters ignited at 0815:30 Houston time, during the 225th orbit. They burned for two minutes and thirty-three seconds and sent *Columbia* on its way back to earth.

At 0844:09 the shuttle flew through the entry interface, at an altitude of 75 miles above the Pacific Ocean. The first effects of the atmosphere could be felt at this altitude. As expected, the growing friction of the increasingly dense layers of atmosphere caused the temperature on the leading edges of the wings to rise to more than 2,460 degrees Fahrenheit in the next six minutes. The heat caused air molecules to form plasma around the shuttle, which the astronauts could see as bright lights outside the windows.

At 270 seconds after the start of reentry, a sensor in the left wing registered a higher temperature than had ever been measured during

Image from a video taken during reentry: Commander Rick Husband (left) and Pilot Willie McCool. *NASA*

this phase of flight in a shuttle mission. This alarming value was, however, recorded only in the orbiter's internal data system and was not reported to the crew or transmitted by telemetry to Houston. Thus the first sign of a looming disaster was hidden.

Under control of the autopilot, the *Columbia* began the first of a series of extended left and right turns. These were to help slowly reduce the sink rate and keep the temperature rise on the shuttle's external hull within limits. At this point the shuttle was flying at 24.5 times the speed of sound.

At 0853:25 the *Columbia* crossed the coast of California at an altitude of 43.5 miles west of San Francisco, still at twenty-three times the speed of sound. Outwardly all appeared to be normal, until at 0854:24, as the shuttle crossed the Nevada border, when the first unusual data appeared on the monitors in Mission Control.

LeRoy Cain, the controller responsible for monitoring the mechanical structure and the life support system (MMACS for maintenance, mechanical, and crew systems), reported to the flight director the loss of four sensors in the hydraulic system of the left wing: "FYI, I've just lost four separate temperature transducers on the left side of the vehicle." Cain asked in response, "OK, is there anything common to them? DSC [discrete signal conditioner] or MDM [multiplexer-demultiplexer] or anything? I mean, you're telling me you lost them all at exactly the same time?" The MMACS stated his information more precisely—that the sensors in the left wing had failed within a few seconds and were independent of one another. "No commonality," answered Cain pensively. After a longer pause the flight director asked the flight controller responsible for flight control and navigation (GNC or guidance and navi-

The control center in Houston. This photo was taken at roughly the time at which the flight controllers lost contact with *Columbia*. *NASA*

gation control) if he had noticed anything unusual. The GNC responded in the negative; everything was normal. When asked again, the MMACS replied that apart from the values from the four sensors in question, all other temperatures were within the normal range.

At 0858 the *Columbia* crossed the border between New Mexico and Texas. About a minute later the MMACS spoke again: "We have just lost tire pressure in both inner and outer left tires." Flight Director Cain instructed CAPCOM to transmit this report to the shuttle crew.

One and a half minutes later, the orbiter was approaching the city of Dallas at an altitude of 37 miles and at eighteen times the speed of sound, when an incomplete radio message was received from the shuttle commander: "Roger ,uh buh . . ." What no one knew then was that this was the last sign of life that was to be heard from the crew of the *Columbia*. At that moment the loss of radio contact bothered no one. They expected nothing different because of the automatic switching of the radio channel from one communications satellite to another at that moment.

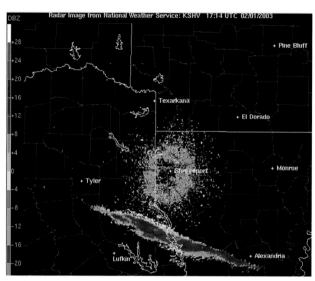

The cloud of wreckage in a photo of the National Weather Service's weather radar at Shreveport, Louisiana. *US National Oceanic and Atmospheric Administration*

On the ground, Flight Director Cain continued to try to make sense of the incoming reports: "And there's no commonality between all these tire pressure instrumentations and the hydraulic return instrumentations." The MMACS confirmed again: "No sir, there's not. We've also lost the nose gear down talkback and the right main gear down talkback."

At that time, eyewitnesses in Texas could already see the orbiter, which was still a bright spot moving slowly through the sky, suddenly surrounded by bright flashes. It was 0900:18 Houston time. Evaluation of the photos and film footage taken at that time later suggested that the space shuttle *Columbia* had obviously broken up at that time. Because of the forty-six-second communications interruption, no one in Houston learned about the catastrophic event.

A short time later the flight controller responsible for instruments and communications (INCO or instrumentation and communications

officer) spoke: "FLIGHT, INCO, I didn't expect, uh, this bad of a hit on comm."

At 09:02:21 a NASA spokesman informed the public on NASA's own radio and television network: "Fourteen minutes to touchdown for Columbia at the Kennedy Space Center. Flight controllers are continuing to stand by to regain communications with the spacecraft . . ." At that time, no one could know that they were waiting in vain.

In the meantime, a dialogue had developed between the flight director and the INCO. Cain: "INCO, we were rolled left last data we had and you were expecting a little bit of ratty comm, but not this long?" Answer: "That's correct, FLIGHT. I expected it to be a little intermittent. And this is pretty solid right here."

Seconds later, Flight Director Cain received the report that the orbiter would be within radio range of the Kennedy Space Center within the next two minutes. He instructed the CAPCOM to call the *Columbia*. The CAPCOM began calling the *Columbia* at regular intervals: "Columbia, Houston, comm check." One minute later the NASA press spokesman made an announcement to this effect: "CAPCOM Charlie Hobaugh calling *Columbia* on a UHF frequency as it approaches the Merritt Island tracking station range in Florida. Twelve and a half minutes to touchdown, according to clocks in Mission Control."

In Houston the flight director and MMACS resumed their conversation. MMACS: "On the tire pressures, we did see them go erratic for a little bit before they went away, so I do believe it's instrumentation." In Mission Control there was still no concrete indication of the dramatic events in the sky over Texas. Instead the attention of the flight director was still concentrated on reestablishing radio contact with the shuttle. CAPCOM continued calling *Columbia* in an increasingly urgent tone: "*Columbia*, Houston, UHF comm check." Flight Director Cain was advised of the lack of success of these efforts: "MILA MILA not reporting any RF at this time." He instead turned to the flight controller responsible for tracking (FDO or flight dynamics officer): "FDO, when are you expecting tracking?" His terse reply: "One minute ago, FLIGHT."

It was 9:12:39 hours, and twenty-eight minutes had passed since the shuttle had flown through the entry window. By that point *Columbia* should have been in a left turn that would have lined it up with Runway 33 of the Kennedy Space Center. The NASA staff and the many guests and family members of the astronauts gathered there should have been able to see the orbiter.

At about that time, a flight controller received a call on his cell phone. He was told that the breakup of the shuttle had just been seen on television. The controller decided to take the news personally to his flight director. He went to Cain's desk and told him that the shuttle had just broken up.

Cain had no choice but to accept the inevitable. He instructed his controllers to lock the control room doors, to stop using their telephones, and to begin carrying out the emergency plans. At many NASA stations, which were directly or indirectly involved with the shuttle flights, as well as at the various supplier companies, documents and data recordings were secured. Within a few minutes the NASA emergency team

was activated, its task to search for wreckage and the remains of the astronauts.

News of the crash of the *Columbia* spread around the globe like wildfire. On the evening of the accident, President George W. Bush gave a speech to the nation in which he mourned the loss of the astronauts.

Immediately after the announcement of the accident, a huge search-and-recovery action was initiated. Pieces of wreckage from the *Columbia* had rained down over an area of more than 1,930 square miles. It borders on a miracle that no one was struck by wreckage, although there were reports of near misses. Herds of cattle in central Texas were startled and bolted. An angler saw a large piece of wreckage fall into a reservoir on the Texas-Louisiana border. Near the small town of Lufkin, a piece of wreckage struck the windshield of a driver's car. Its female driver was not injured.

Eastern Texas was declared a disaster area. NASA warned the population not to touch the wreckage, since it feared that poisonous fuels and other toxic chemicals might be present. Explosive charges on the shuttle might well have survived the crash and posed a po-

Moment of truth: Flight Director LeRoy Cain (right) realizes that the shuttle has been lost. *NASA*

A gruesome piece of wreckage: the helmet of one of the *Columbia* astronauts. *NASA*

One of the most important finds was the modular auxiliary data system recorder. Data from more than 800 measuring sites saved on the device's magnetic tape were of inestimable value to the investigators. Together with recordings of telemetry data and film and photos of the shuttle breakup taken from the ground, it formed the basis for understanding what had happened high over Texas that morning.

Soon, various theories about the cause of the crash were under discussion: possibly the heat shield had been damaged or the autopilot, which according to the flight plan was in control of the shuttle at the time of the accident, could have failed. Mechanical failure of the structure of the shuttle, which after all was more than twenty years old, seemed another possible answer, as did a fire onboard or collision with space garbage. In the initial excitement, a terrorist attack was also not ruled out; after all, an Israeli had been part of the doomed shuttle's crew.

A few hours after the accident, NASA administrator Sean O'Keefe summoned an external investigation committee, the Columbia Accident Investigation Board (CAIB). Retired admiral Harold Gehman was named to head the board. The committee published its findings in a 248-page report on August 26, 2003.

The pieces of wreckage were gathered in a hangar at the Kennedy Space Center. The examination of some pieces of wreckage quickly revealed that the accident must have begun in the leading edge of the left wing. This theory was also supported by the first unusual sensor readings from that area. The investigators came to an awful suspicion.

In the investigating committee's final report, the sequence of events was listed in min-

tential threat. The space authority laid claim to all remains, which they urgently needed to determine the sequence of events of the accident and its cause. Later, the keeping of wreckage was even made a criminal offense.

Initially the search-and-recovery effort was carried out by local police, firefighters, and disaster control agencies. Later the search teams were augmented by units of the military and National Guard and by forestry and environmental authorities. At times there were as many as 4,000 people simultaneously involved in the search, with a total of 25,000 people altogether. Seven airplanes and thirty-seven helicopters supported the search effort from the air. Even a U-2 spy plane was brought in to take aerial photographs. Two people were killed in a helicopter crash, and three others were injured. The total cost of the search effort is estimated to have exceeded 350 million US dollars. Altogether more than 82,000 pieces of wreckage with a total weight of 42.4 tons were recovered. This was equal to about 38 percent of the shuttle.

Some of the more than 82,000 pieces of wreckage from the *Columbia*. *NASA*

ute detail. First, sensors had registered a higher dynamic load in the forward part of the left wing, followed by unusually high temperatures in that area. In the minutes that followed, the temperature rise spread to the entire left wing. At about that time, witnesses on the ground observed pieces start to separate from the orbiter. The plasma contrails behind the *Columbia* became irregular. As the *Columbia* passed over Nevada, bright flashes were visible around the shuttle. The orbiter continued on its path over the states of Utah, Arizona, and New Mexico to Texas. More and more witnesses reported seeing pieces separate and fall to earth.

At the same time, the pressure indication for the tires of the left main undercarriage dropped. A major control movement by the orbiter's rudder suggested altered flow conditions on the wings. The problem appeared to have begun in the left wing.

In Houston, telemetry data were received from various sensors, signaling that the left main undercarriage had extended. Other sensors, however, indicated the opposite. The last muffled radio message from the crew was recorded. Telemetry transmissions ended at about the same time. Just thirty seconds later, observers saw the complete breakup of the shuttle. The crew compartment, which had obviously been intact until then, broke up seconds later. A literal shower of wreckage then fell over Texas and the bordering states.

The investigators soon turned their attention to an incident that had occurred sixteen days before the accident; namely, the shuttle launch. At first it appeared that the launch had been normal, without any unusual events. Nothing out of the ordinary was seen on live pictures and video recordings. The next day, however, engineers and project managers had

In the sky over Texas, the space shuttle *Columbia* can clearly be seen to have broken into several pieces, which are now burning up in the atmosphere. *NASA*

1,650 miles per hour. Detailed analysis of the images led to the conclusion that the piece of insulation that had fallen off was about 12 inches long and 15.75 inches wide. The impact with the wing must have taken place at a relative velocity of 530 miles per hour.

That same day, the photo evaluators requested project manager Linda Ham to contact the Defense Department and ask for pictures of the shuttle in orbit. It was known within NASA that the department had the technical capability to obtain high-resolution images of objects in orbit. The military used ground-supported systems and reconnaissance satellites with the necessary configuration. It was hoped that such photos might reveal damage caused by the impact. This request for help was the first attempt to obtain supporting photographic material from outside NASA.

available high-resolution images from high-speed cameras for the routine video analysis of the launch. In these images, at 81.7 seconds after liftoff they could clearly see a piece of insulation from the external tank separate and fall off. The foamlike piece separated from the area of the bipod that connected the external tank with the nose of the shuttle. It flew along the orbiter's fuselage and 0.161 seconds later struck the leading edge of the left wing. There it burst asunder into a cloud of tiny pieces. At the time of the incident, the orbiter was at an altitude of about 65,000 feet at a speed of

At the same time, a working group of engineers and managers and representatives of subcontractors had been formed to investigate the possible effects of the impact. The group began its work on the fifth of the planned sixteen mission days, and it also requested supporting pho-

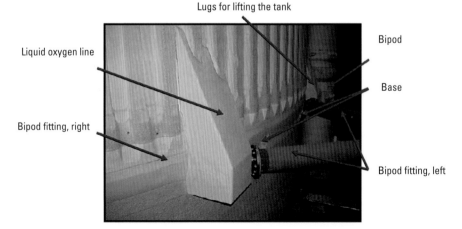

Close-up of the forward bracket of the external tank. It was from this area that the piece of foam that damaged *Columbia*'s left wing came. *NASA/Woydt*

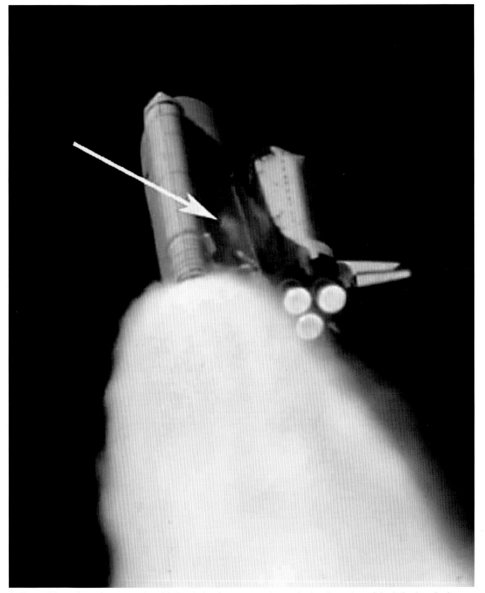

A piece of foam from the external tank's insulation bursts asunder on the leading edge of the left wing. At that time no one suspected that the fate of the crew was decided at that moment. *NASA*

tographic material from the military. Lacking such photos, the working group had been forced to rely on computer simulations, whose exactness was quite questionable, however. On the basis of these simulations, the working group concluded that the impact by the piece of foam probably could not have seriously damaged the leading edge of the wing. This was, however, based on the proviso that the angle of impact had been less than 21 degrees, which seemed to have been the case. This would have been expected to create local warming in the affected area of the orbiter, if at all. The full extent of the assumptions and the potential inaccuracy of the estimates by mission management were never revealed, however.

On the seventh day of the mission the working group made one final effort to obtain supporting photographic material. This time its members took the direct route rather than the official one, going directly to the technical offices to ask for their help. The program manager was aware of this latest request, and he confirmed that the request had been made purely out of technical interest. The requested photographic material was not judged to be critical to the investigation, and the request was again denied. The matter was thus closed.

At a later hearing, Linda Ham, chairman of the mission management team, justified her decision by declaring that they could not have done anything in an emergency anyway. There were no plans for the crew repairing the shuttle in orbit, and thus it could not have been carried out. Because of the shuttle's orbital parameters it would also have been impossible to head for the International Space Station to use it as a quasi life raft. A rushed rescue mission by a second space shuttle would have been conceivable in theory, but it would have involved major risks for those involved.

Notwithstanding the discussions on the ground, on January 23, the eighth day of the mission, Commander Rick Husband was told about the impact with the piece of insulation. He was not told as a warning about possible damage and its potential consequences, but for

another reason. An email from Flight Director Steve Stitch said, "There is one item that I would like to make you aware of for the upcoming PAO event on Blue FD 10 and for future PAO events later in the mission. This item is not even worth mentioning other than wanting to make sure that you are not surprised by it in a question from a reporter." He concluded with these words: "Experts have reviewed the high-speed photography and there is no concern for RCC or tile damage. We have seen this same phenomenon on several other flights and there is absolutely no concern for entry." The next day, Commander Husband thanked Stitch for the email, without going into its contents. One day later the flight control center even sent video images of the impact event up to *Columbia*. The erroneous assessment on the part of the NASA manager would have fatal consequences on the day of the recovery.

Although the accident investigators were certain they had found the cause of the accident and understood the course of events, they tried to confirm their theory through experimentation. For this purpose they used a faithful mock-up of a shuttle wing. Using a special pressurized gun, a piece of foam of the same condition and roughly the same dimensions and estimated weight as the piece that fell off on the day of the launch was fired at the mock-up. Several tests were carried out, and in one of them the piece of foam did in fact did make a large hole in the heat-shielding elements of the leading edge of the wing.

Now it was all clear. A piece of foam from the tank insulation had struck the leading edge of the left wing, seriously damaging the orbiter's heat shield. The damage was in the precise area that would have stood up to the highest temperatures during reentry. Hot plasma, with a temperature in excess of 1,800 degrees Fahrenheit, was forced through the hole and into the interior of the wing.

The plasma initially severed electrical and hydraulic lines and then destroyed the wing support structure with its ribs and struts. The aerodynamic characteristics of the wing and thus the entire shuttle changed very quickly, and the crew lost control of the *Columbia*.

It took just forty seconds for the orbiter to be completely torn apart by the tremendous buildup of aerodynamic forces. The crew compartment obviously remained intact initially, until it was destroyed by the effects of air friction twenty seconds later.

The accident investigation was concluded. The Columbia Accident Investigation Board compiled its findings in a report that was submitted to the president and to Congress.

NASA reacted to the second accident involving a space shuttle with the somewhat unwieldy-sounding title *Space Shuttle Return to Flight Implementation Plan*. The realization of this plan was supposed to address the points that the Columbia Accident Investigation Board had admonished as critical, questionable, or at least in need of improvement in its investigation results. For example, far-reaching changes in the NASA structural organization were initiated, and technical improvements were also made to the orbiter. The external tank was redesigned to prevent to the degree possible the separation of foam insulation and the creation of dangerous fragments in the future.

Another measure that was taken was the creation of the Spacecraft Crew Survival Integrated Investigation Team (SCSIIT). The group's task was to carry out a thorough reanalysis of

A piece of foam insulation that was fired from a pressurized gun at a replica of the leading edge in tests to determine the cause of the accident.

The pressurized gun from which a piece of foam was fired at a true-to-life piece of the wing (red circle) from a space shuttle. *NASA*

The shocking result of the experiment: a large hole in the leading edge of the wing. Traces of an earlier test can be seen to the left—there the impact caused no damage. *NASA*

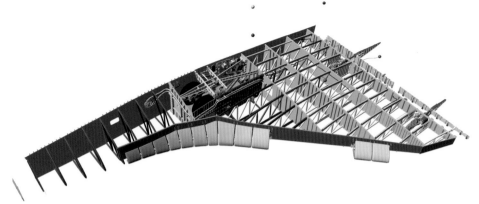

Cutaway drawing of the space shuttle's left wing. *NASA*

the *Columbia* accident, this time paying special attention to aspects that had led to the death of the crew. The team had to evaluate incidents that lay far outside the familiar framework of aircraft accident investigations. This proved extremely difficult. Its findings were based on video evidence and examinations of the physical remains of the crew.

Among the working group's most important findings was that not all of the seven *Columbia* crewmen had followed the safety regulations of the reentry phase. Among other things, these rules called on the crew to wear their

Beschädigter Bereich
Bereiche betroffener Sensoren

Diagram of the areas damaged by the impact of the foam, and the positions of the sensors that failed during reentry, providing the first indication that something was wrong. *NASA*

spacesuits, including gloves and helmets. Their visors could remain open, however. In contrast to the regulations, however, three of the astronauts had not put on their gloves, and one had not even put on his helmet.

During the *Columbia* accident, the first event with possibly fatal consequences had been the loss of pressurization in the crew cabin. Between the loss of the ability to control the shuttle and its breakup with resulting pressure loss, the astronauts had about forty seconds in which to close the visors of their helmets. For inexplicable reasons, however, they failed to do this. The sudden loss of oxygen must therefore have resulted in all seven crew members losing consciousness almost immediately. Furthermore, the centrifugal forces to which the astronauts were exposed in the wildly spinning cabin would have been sufficient to inflict fatal injuries. The rate of rotation probably reached one revolution every ten seconds. Under this load the astronauts' upper body belts would have failed, which would have left just their lap belts holding them down in their seats. The centrifugal forces were not so great, however, that a relatively well-trained person could not

Press conference by the investigating committee looking into the *Columbia* disaster. *NASA*

have resisted them. The helmets were not, however, optimized to the shapes of their wearers' heads, like motorcycle helmets for example. This would have further increased the probability of fatal injuries.

The third event with potentially fatal consequences would have been the breakup of the cabin. At the moment of its destruction, the shuttle was flying at 11,500 miles per hour at an altitude of 34 miles. When the bodies of the astronauts were thrown from the disintegrating crew compartment, they would have been abruptly exposed to tremendous aerodynamic forces and the friction heat of reentry. Sharp-edged pieces of metal flying about would have posed an additional threat. The astronauts' spacesuits would have offered no protection under these extreme conditions.

Whether or not the astronauts had been killed by the first or second events, the third and also the most serious event would in any case have meant death. The astronauts may have been conscious and aware that there was absolutely nothing they could do for the first forty seconds of the accident. Considering that, the abrupt pressure loss and the resulting unconsciousness must have been a blessing. Death occurred during the breakup of the crew cabin at the latest, and the astronauts did not have to live through the more than thirty-minute fall until their bodies struck the earth.

The physical remains of the astronauts were turned over to their families. Commander Rick Husband was buried in Llano Cemetery in his home town of Amarillo, Texas. Pilot William McCool was buried in an unmarked grave in his

Residents of the town of Hemphill in Texas placed this memorial at the spot where the remains of a crew member were found.

wife's birthplace in Anacortes, Washington. The graves of David Brown, Michael Anderson, and Laurel Clark are in Arlington National Cemetery in Arlington, Virginia. The physical remains of Kalpana Chawla were cremated and her ashes were spread in a national park in Utah. The body of Ilan Ramon was flown to Israel, where he was buried in a small village named Nahalal, west of Nazareth.

After a pause of two and a half years, the space shuttle program resumed on July 26, 2005, with mission STS-114 by the orbiter Discovery.

Sequence of Events

Time (Houston Local Time)	
8:10:00	*Columbia* receives authorization to ignite the engines for the reentry burn maneuver.
8:15:30	Burn maneuver begins.
8:44:09	*Columbia* enters the earth's atmosphere.
8:48:39	Sensors register a higher load on the forward part of the left wing structure than expected.
8:48:59	Sensors inside the left wing register an unusual rise in temperature.
8:49:32	*Columbia* begins a planned roll maneuver to the right.
8:49:49	More sensors signal an unusual temperature rise.
8:50:53	As expected, *Columbia* begins the ten-minute phase of highest temperatures caused by the increasing friction with the thickening atmosphere.
8:51:14	More sensors in the left wing register an unusual rise in temperature.
8:52:00	Temperature on the leading edge of the left wing is 2,642 degrees F., as expected.
8:52:16	First parts of the structure inside the left wing burn through.
8:52:51	First temperature sensor inside the left wing fails.
8:52:59	More temperature sensors inside the left wing fail.
8:53:26	*Columbia* crosses the coast of California west of San Francisco.
8:53:44	First eyewitness reports of pieces separating from *Columbia*. Additional reports are received in rapid succession.
8:54:11	Roll is reversed to the left.
8:54:24	The flight director in Houston receives the first reports about problems with *Columbia's* sensors, the first indication that the reentry is not proceeding normally.
8:54:25	*Columbia* crosses the border with Nevada; seconds later, a bright flash of light is seen.
8:54:33	A bright flash of light is seen, leaving behind recognizable irregularities in the shuttle's glowing plasma trail.
8:55:00	The temperature on the leading edges of both wings reaches 3,000 degrees F., as expected.

8:55:32	*Columbia* crosses the Utah border.
8:55:52	*Columbia* crosses the Arizona border.
8:55:58	Very bright pieces of wreckage are seen separating from the plasma trail.
8:56:30	*Columbia* begins a roll movement to the left.
8:56:45	*Columbia* crosses the Arizona border.
8:56:55	Roll maneuver completed.
8:57:19	Tire pressure on the outer side of the left main undercarriage drops.
8:57:24	Tire pressure on the inner side of the left main undercarriage drops.
8:58:00	Temperature on the leading edges of the wings has fallen to 2,876 degrees F., as expected.
8:58:03	The autopilot begins elevon trim motions.
8:58:20	*Columbia* crosses the Texas border; the first pieces of wreckage strike the earth.
8:58:38	Pressure indication in the outer tires of the left main undercarriage is lost.
8:58:48	Tire temperature indicators for the outer side of the left main undercarriage are lost; another sensor signals no change in the condition of the undercarriage.
8:59:06	A sensor signals that the left main undercarriage has lowered; another sensor signals no change in the condition of the undercarriage.
8:59:15	The flight director is informed that the pressure indicators for both tires of the left main undercarriage have been lost.
8:59:32	A last garbled radio message from Commander Rick Husband is received; telemetry data are lost.
9:00:18	Video images show the breakup of the shuttle.
9:00:19	Data recording on the orbiter ends.
9:00:21	The fuselage of the *Columbia* breaks up.
9:16:00	The planned time for landing has passed; emergency plans are implemented.
9:30:00	The Columbia Accident Investigation Board is convened.
9:30:00	NASA administrator Sean O'Keefe meets with families of the Columbia crew.
13:00:00	The NASA administrator personally informs the president and other important government offices.

The External Tank of the Space Shuttle

The external fuel tank represented the largest single component of a space shuttle. It consisted of two separate tanks, in which the deeply cooled components of the fuel for the three main engines were maintained in a liquid state. The upper of the two tanks held 145,295 gallons of oxygen at –297 degrees Fahrenheit. The roughly 369,840 gallons of hydrogen in the lower tank was even colder, –423 degrees Fahrenheit. The two individual tanks were joined by an intertank.

The tank was built by Lockheed Martin and was made largely of aluminum alloy. It was 154 feet long and 27.5 feet in diameter. The intertank between the two fuel tanks, the central components, was 22.6 feet long.

The two solid-fuel rockets, which provided 80 percent of the shuttle's total thrust during launch, were attached to this intertank at the upper end of the tank by explosive bolts. The lower part of the booster was attached to the lower part of the hydrogen tank in a similar way.

The shuttle itself was attached to the external tank by two braces near the engines. Through this connection ran electric cables and the fuel lines for the hydrogen and oxygen to the shuttle's engines. At the front of the shuttle a V-shaped brace joined the orbiter and the external tank.

The external tank was covered in a rigid foam material, which was sprayed onto its

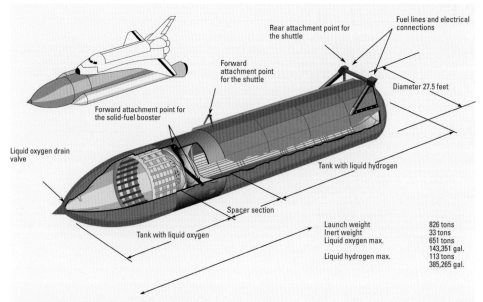

Rear attachment point for the shuttle

Fuel lines and electrical connections

Forward attachment point for the shuttle

Diameter 27.5 feet

Forward attachment point for the solid-fuel booster

Liquid oxygen drain valve

Tank with liquid hydrogen

Spacer section

Tank with liquid oxygen

Launch weight	826 tons
Inert weight	33 tons
Liquid oxygen max.	651 tons
	143,351 gal.
Liquid hydrogen max.	113 tons
	385,265 gal.

Schematic of the external tank. *Malyszkz/Woydt*

hull. The coating provided insulation, drastically reducing the loss of ice-cold fuel due to evaporation. It also reduced the formation of hoarfrost and ice on external mechanical components such as the attachment brackets. In the area of the intertank the insulation was up to an inch thick.

After launch, the solid-fuel rockets burned out at an altitude of about 30 miles. They were then jettisoned and the shuttle's three main engines with their burn duration of eight and a half minutes delivered the orbiter to a 68-mile-high orbit around the earth. The shuttle's speed was then about 17,200 miles per hour. The empty tank was jettisoned, entered the atmosphere, and largely burned up. A few fragments, after having flown halfway around the earth, fell into the

ocean about 11,500 miles from the launch site at Kennedy Space Center in Florida. Depending on the trajectory, the impact site could be in the Indian or the Pacific Oceans.

Power to achieve an ultimate orbit was provided by the two engines of the shuttle's orbital maneuvering system (OMS).

The huge external tank of the space shuttle. The surface structure of the foam insulation is clearly visible. *NASA*

IS IT POSSIBLE TO DROWN IN SPACE? THE PARMITANO INCIDENT

The International Space Station (ISS) is a joint project involving NASA, the Russian space agency (Roskosmos), the European ESA (European Space Agency), the Canadian CSA (Canadian Space Agency), and the Japanese JAXA (Japan Aerospace Exploration Agency). As a result of an agreement signed in 1988, the European nations of Belgium, Denmark, Germany, France, Italy, the Netherlands, Norway, Sweden, Switzerland, and Spain became involved in the construction and operation of the ISS.

The ISS has been in space since November 29, 1998. On that day the first module of the space station, the Russian propulsion and freight module Zarya, was placed in orbit by a Proton rocket. Just fourteen days later a space shuttle crew added the first American module. The individual components of the modularly built ISS were gradually transported into space by using Russian Proton and Soyuz booster rockets and the American space shuttle fleet and were assembled there. The first long-term crew, designated ISS Expedition 1, was finally able to move into the station on November 2, 2000. Since then, the ISS, also called "humanity's outpost in space," has been constantly occupied as it circles the earth every ninety-two minutes at an altitude of 250 miles.

Since its creation, the space station has been visited by many manned and unmanned spacecraft. Supplies of equipment, food, water, and scientific equipment are in part delivered by unmanned freighters, such as the Russian Progress space freighter or the European ATV, in some cases supported by shuttle flights.

Russian Soyuz capsules and, until the shuttle fleet was grounded, American spacecraft have taken many space travelers to the space station as long-term crews or for more or less short visits.

The first twelve long-term crews were composed exclusively of Russian cosmonauts or American astronauts. Since ISS Expedition 13, however, astronauts of the CSA, ESA, and JAXA have also made lengthy visits.

On May 29, 2009, the nineteenth and twentieth crews met on the International Space Station. Since that time the station has been permanently manned by six space travelers. In total, more than 200 people have spent time on the ISS.

A number of astronauts have carried out extravehicular activities, or EVAs, to expand the station and to carry out scientific experiments. Such EVAs are physically and psychologically very demanding and remain as risky today as they were in the early days of space travel.

A serious incident during one such EVA occurred on February 10, 2001, during the first of

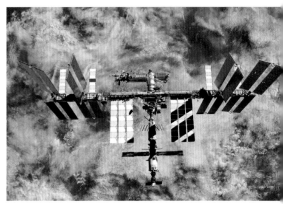

The International Space Station in its current configuration. *NASA*

The Spacesuits Worn by the ISS Astronauts

The spacesuit currently used on the International Space Station for extravehicular activities is, like all its predecessors, a system independent of other supply installations, which provides its wearer with life support functions, protection against the inhospitable conditions of space, mobility, and communications equipment. The current version has been in use, with some improvements, since 1998.

Because of its extensive functions, the spacesuit is called an extravehicular mobility unit or EMU. On earth it weighs an impressive 319 pounds.

The EMU consists of two main components. The internal pressure unit offers its wearer a small, closed world with breathable air and acceptable temperatures. Under normal conditions the atmospheric pressure inside it is 0.3 atmospheres. Its equipment includes a 1.0- to 1.6-quart pouch with drinks for the astronauts, which is worn on the chest. A tube leading into the helmet allows the wearer to suck liquid from the pouch.

The shell of the pressure unit is about 0.60 inches thick and consists of two layers. The inner layer of laminated nylon is gas impermeable and prevents breathing air from escaping into the vacuum of space. The outer layer consists of aluminum-metalized polyester. It provides heat insulation and protects against the extreme temperatures of space.

The EMU's second main component is the portable life support system, or PLSS, which is worn on the back like a rucksack. Part of the PLSS is the secondary oxygen package, or SOP, for emergencies. The PLSS also includes two high-pressure tanks for oxygen and a system for recycling breathing air. A cooling system is integrated for temperature regulation. Pumps force warm water from two fluid reservoirs through special astronaut underclothing, which is infused with a network of fine tubes. Ventilators ensure the continuous exchange of used breathing air.

The spacesuit's life support system can keep an astronaut alive for up to seven hours under light to moderate physical loads, and up to an hour longer in an emergency.

In addition to life support equipment, the PLSS also includes equipment for communication with the other astronauts on the ISS

Helmet lamp
Headphone and microphone cap
Helmet
Glove
Display and control module
Sight for extravehicular activities
Boot

The spacesuit (extravehicular mobility unit) worn by ISS astronauts. *NASA* and flight controllers on the ground. It transmits not only voice traffic, but also telemetry data about the condition of the spacesuit and vital parameters of its wearer.

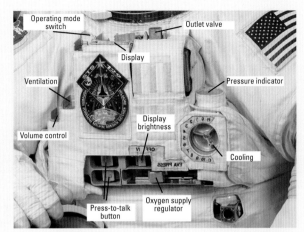

The internal components of the EMU are surrounded by a semirigid shell that shields it from the extreme cold of space and the heat of direct sunlight. It also shields the wearer against micrometeorites and

The spacesuit's display and control module, or DCM. *NASA/Woydt*

protects the sensitive inner life of the spacesuit against unintentional damage during work in space.

Outside the shell of the spacesuit, the astronaut wears the display and control module, or DCM, on his chest. With its help he can monitor the levels of the suit's gas and liquid tanks, operate the communications system, and select temperature, the composition of the breathing air, and other life support parameters to suit his desires.

Also part of the spacesuit in the broader sense is an emergency system with the somewhat cumbersome title of the Simplified Aid for EVA Rescue system, or SAFER. It is a sort of rack attached to the spacesuit's life support system and is designed to be a self-help emergency system. With its help the astronaut can return to his space vehicle should his security line break or if he drifts into space for some other reason. He operates twelve small nitrogen jets, with which he can maneuver or make changes of direction. The rescue system has been available since 1994, but fortunately it has never had to be used in a real emergency.

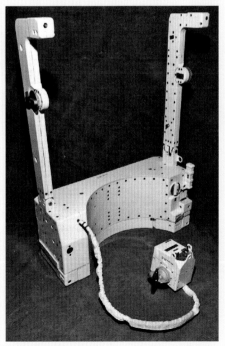

The SAFER (Simplified Aid for EVA Rescue) system. *NASA*

three extravehicular activities during the space shuttle *Atlantis's* STS-98 mission to the ISS. The American astronauts Robert Curbeam and Thomas Jones had the task of attaching the newly installed Destiny laboratory module to the ISS's cooling system with special lines. A defective valve caused approximately 5 percent of the space station's coolant, extremely poisonous ammonia gas, to escape into space. While Curbeam was attempting to close the stuck valve, some of the ammonia, which had frozen in the cold of space, settled onto his spacesuit. His helmet and part of his suit were finally covered in a layer of ammonia crystals up to one inch thick.

To avoid contamination of the space station, ground control instructed the astronauts to remain in the sunlight during the next pass over the daylight side of the earth. This was supposed to cause the crystals to vaporize so that they would not be carried into the interior of the ISS by the astronauts.

The long-term crew of ISS Expedition 36. Bottom, from left to right: Pavel Vinogradov, Karen Nyberg, and Alexander Misurkin. Top row, left to right: Luca Parmitano, Chris Cassidy, and Fyodor Yurchikhin. *NASA*

After Curbeam and his colleague Jones had returned to the space shuttle's airlock, pressure equalization was carried out. Afterward the air in the airlock was vented once again, to remove the last of the ammonia crystals from the air.

After the airlock's inner hatch was opened, the entire crew of five wore gas masks for half an hour to be certain. Measurements revealed no contamination of the atmosphere inside the shuttle, and so the incident ended well. No one was injured and the damage to the space station's cooling system was repaired during the next EVA.

On May 28, 2013, a Soyuz TMA spacecraft headed into space. Its three-man crew under the command of veteran Russian cosmonaut Fyodor Yurchikhin docked at the International Space Station one day later. Its crew included female astronaut and flight engineer Karen Nyberg, who was on her second spaceflight. With them was Italian space newcomer Luca Parmitano, representing the ESA. At the station the three newcomers met the other three people living on the station, who had been there for six weeks. They were the ranking Russian commander Pavel Vinogradov, his countryman Alexander Misurkin, and the American Christopher "Chris" Cassidy. With their arrival, Yurchikhin, Nyberg, and Parmitano became members of ISS Expedition 36.

At thirty-seven years of age, the Italian was the youngest man so far assigned to a long-term ISS crew. The ESA astronaut was married and the father of two daughters. Before his space mission, Parmitano, whose hobbies included diving, had served as a test pilot with the Italian air force. He had demonstrated great courage and flying skill when he succeeded in landing a badly damaged military jet. On May 11, 2005, a large bird, probably a stork, had col-

Italian ESA astronaut Luca Parmitano. *NASA*

Luca Parmitano in his spacesuit during preparations for the first extravehicular activity. *NASA*

lided with his aircraft, shattering the plane's windscreen. Even worse was the fact that the head-up display, which displayed flight information and data about the condition of the aircraft in the pilot's field of view, was no longer working. Radio communication was almost impossible due to the noise in the cockpit caused by the shattered windscreen. Despite everything, Parmitano kept his nerve, and he decided not to eject from his aircraft. With the help of a squadron mate flying beside him, who used hand signals to give him instructions and assistance, he was able to safely land his badly damaged aircraft. For this extraordinary accomplishment he was awarded the Italian air force's Silver Medal of Honor.

In space, Cassidy and Parmitano completed their first joint extravehicular activity on July 9, 2013. It was Cassidy's fifth EVA, while Parmitano became the first Italian ever to move about freely in space. The two astronauts worked on the space station's outer hull. They laid cables,

Luca Parmitano onboard the International Space Station during a technology experiment. *NASA*

collected samples from various material experiments, and installed cameras and foot rests for future extravehicular activities.

After somewhat more than six hours, the two returned through the ISS's airlock. When removing Parmitano's spacesuit the astronauts

Sequence of Events during the Extravehicular Activity on July 16, 2013

Minutes	
00:00:00	Extravehicular activity begins.
00:44:00	Luca Parmitano reports: "I feel a lot of water on the back of my head, but I don't think it is from my bag."
00:47:00	The controller responsible for monitoring the extravehicular activity suspects that the water is coming from a vent port.
00:48:00	Ground control asks Parmitano if he can identify the cause for the loss of water and if the amount of water in his helmet is increasing.
00:49:00	Parmitano reports: "I still feel it and I cannot tell you the source. My only guess is that it came out of my bag and then found its way over there in the back, but I don't have any water in the front of the helmet."
00:54:00	Parmitano reports: "The leak is not from the water bag, and it is increasing." He receives no answer.
00:54:30	The controller responsible for monitoring the extravehicular activity asks spacesuit experts if the EVA should be terminated.
00:55:00	Parmitano drinks the rest of the water in his drink bag.
00:56:00	Chris Cassidy looks into Parmitano's helmet. He estimates that about half a liter of water has collected in the back of the helmet.
00:57:00	The flight director discusses with the controller responsible for monitoring the extravehicular activity the fact that Parmitano's water bag is now empty. Both therefore suspect the water bag as the most likely source of the liquid in his helmet.
01:00:00	Parmitano declares: "I'm thinking that it might not be the water bag." He receives no answer.
01:00:30	Cassidy asks whether the liquid in the helmet could be sweat or urine.
01:01:00	The controller responsible for monitoring the extravehicular activity discusses with colleagues whether Cassidy's suspicion could be correct.
01:03:00	Parmitano asks if his underwear could be the cause for the buildup of water.
01:05:00	Parmitano reports that most of the water is in the area behind the back of his head.
01:08:00	Ground control instructs the astronauts to terminate the extravehicular activity.
01:09:00	Parmitano makes his way to the airlock.
01:11:00	Radio communications with Parmitano start to become irregular.
01:12:00	The ISS enters the earth's shadow, and it suddenly becomes dark.
01:29:00	Both astronauts are in the airlock. The outer hatch is closed.
01:32:00	Pressure equalization begins at normal rate.
01:38:00	The inner airlock hatch is opened.
01:41:00	Parmitano's colleagues help him remove his helmet.

discovered that a considerable amount of liquid—they estimated about one half to one quart—had collected in his helmet. Chris Cassidy had not noticed any liquid in the Italian's helmet during the EVA or during the pressure equalization in the airlock. The flight controllers in Houston concluded that the liquid must have gotten inside his helmet only inside the station. They did not pursue the matter further. Their only reaction was to advise the astronauts to use a new drink bag in the spacesuit in question during the next EVA. A dangerous carelessness, as events a few days later would show.

In the early days of space travel, NASA still considered extravehicular activities high-risk activities. More than a few experts believed that the plan to carry out three or even four EVAs per shuttle mission was impossible for this reason. This proved to be a miscalculation. In the first decade of this century, extravehicular activities apparently became routine. Looking back, it seems astonishing that there never was a really serious incident during the more than 1,000 hours astronauts spent assembling the ISS in space.

Luca Parmitano during the extravehicular activity on July 16, 2013. *NASA*

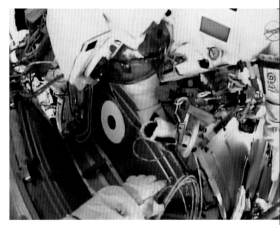

During the extravehicular activity, Chris Cassidy tried to look into his colleague's helmet. He was in fact able to make out liquid floating about inside. *NASA-TV*

A week after their first extravehicular activity, Cassidy and Parmitano prepared for a second spacewalk. With the help of their colleagues, the two astronauts donned their spacesuits—the same ones they had worn during the previous EVA. They released the air from the airlock and at 13:57 Central European summer time they made their way to their workplaces on the space station's outer hull. The duration of the work had been set at six and a half hours. No one, neither on the ground nor in space, suspected what a dramatic turn this seemingly routine mission would take.

Chris Cassidy began by resuming work on the space station's so-called Z1 boom, which had been started a week earlier. Parmitano worked on the electrical wiring on the outer hull of the ISS. The work went well, and after

forty minutes both astronauts were clearly ahead of schedule.

At that time, Parmitano made an unsettling discovery. He was in a confined position where three cylindrical modules came together, when he felt moisture in the neck area of his helmet. Parmitano reported his finding to Shane Kimbrough in Mission Control. "I feel water on the back of my head, but I don't think it's coming from my drinking bag." Unimpressed, the astronaut initially continued his work.

In Houston the controllers were alarmed. Karina Eversley, who was responsible for monitoring the EVA, contacted NASA spacesuit experts and specialists from the manufacturer to localize possible sources of liquid. Their first guess was that an outlet valve inside the spacesuit was leaking.

While the two astronauts waited for instructions from Houston regarding how to deal with the situation, Luca Parmitano could not rid himself of the impression that the amount of liquid in his helmet was increasing. He conveyed his impression to ground control and his colleagues on the ISS: "The leak is not in the drinking bag, and it is becoming larger." Soon afterward he declared more forcefully, "I don't believe that it's the drinking bag."

In space, Chris Cassidy moved to his colleague, hoping that he could see something of the liquid in his helmet. He tried to localize the source of the water by tapping his hand against Parmitano's helmet visor. In fact he could see drops floating about inside the helmet.

Parmitano energetically opposed Cassidy's suggestion that it might be drops of sweat or urine. When Houston asked if the amount of water was increasing, he replied, "It is difficult to say, but it feels like a lot of water." Finally the Italian succeeded in catching a few of the drops floating in front of his face in his mouth. They tasted unpleasantly metallic and definitely not like drinking water.

Houston finally responded after a lengthy discussion among the Mission Control flight controllers. Parmitano and Cassidy were instructed to terminate the EVA. For Parmitano, this meant immediately returning to the space station's airlock. Meanwhile, Cassidy was to collect tools and other equipment and then follow his colleague. An immediate termination of the EVA as prescribed for an emergency situation did not seem necessary to the flight controllers at that time. In that case the two astronauts would have left everything where it was, and immediately returned through the airlock to the ISS.

Parmitano set out to cover the roughly 100 feet to the airlock, while Cassidy stayed behind. Initially the Italian's return was problem free. Then, however, he was forced to change his position to get past an antenna. His feet were now pointing away from the station in the direction of space.

At that moment the situation took a dramatic turn. A large drop of liquid appeared before Parmitano's face. Seconds later it was over his eyes and, what was much worse, in front of his nose. The astronaut had difficulty seeing and could breathe only through his mouth, since otherwise there would be the likelihood of getting water in his lungs.

To make matters worse, at that moment the ISS entered the earth's shadow, complicating the situation even further. It suddenly became pitch black. Normally this would not have been a problem. Every spacesuit was equipped with helmet lamps, which gave a limited but adequate field of view. With the drops of water floating in the helmet and the water in his eyes, from one minute to another Parmitano was almost blind. At the same time he discovered that communications both with Cassidy and ground control were becoming increasingly unreliable. There were frequent interruptions in radio signals. The others could scarcely hear him, and he could almost not hear the others. Parmitano realized that he was now on his own.

The Italian showed strong nerves in this life-threatening situation. Using the hand grips, he felt his way along the outer hull, and with the help of his safety line he pulled himself toward the saving airlock. After seemingly endless minutes, he finally reached it and floated inside.

Inside the space station, the other four members of the ISS crew—Nyberg, Vinogradov, Yurchikhin, and Misurkin—waited impatiently next to the airlock for the return of their two colleagues. They could see Parmitano through a small porthole, but there was no way for them to get to him and render assistance.

Now Cassidy also had to return to the airlock. After his arrival he closed the outer hatch as quickly as possible and began pressure equalization.

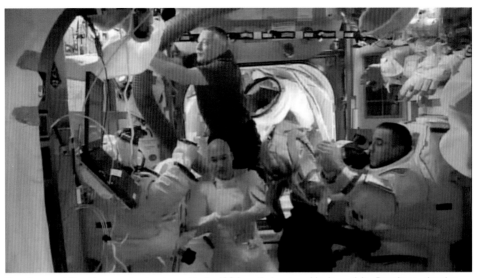

Parmitano and Cassidy prematurely returned to the ISS from their extravehicular activity. Parmitano, showing the effects of what he had just experienced, has already removed his spacesuit. Karen Nyberg is helping Cassidy out of his. *NASA-TV*

By that time Parmitano was completely dependent on the others for help. He could not see or hear anything. He tried to move as little as possible, to keep the floating liquid away from his face. At any moment he might breathe in the water through his nose or mouth. That would have meant death; the Italian would have suffocated miserably.

While Parmitano waited for his colleague's return, he had worked out a desperate plan. If the liquid got into his windpipe and lungs, he would immediately open his helmet no matter how far the pressure equalization had proceeded. The Italian speculated that at worst the insufficient air pressure would render him unconscious, but that would not have been as bad as drowning in his spacesuit.

At ground control in Houston the tension rose from second to second. It had been minutes since the flight controllers had heard anything about the astronaut in danger. Duty Flight Director David Korth instructed the astronauts on the

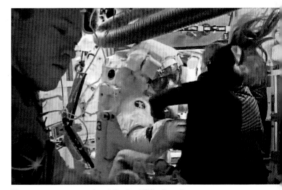

The astronauts are back in the ISS after the extravehicular activity. The crew tries to remove Luca Parmitano's helmet as quickly as possible. *NASA-TV*

International Space Station to carry out the pressure equalization as quickly as possible, even if this meant doing so faster than the rules prescribed. In the worst case, Parmitano's and Cassidy's eardrums would have been damaged. Korth was of course prepared to accept this insignificant risk to save the life of the Italian.

Cooling underwear worn by the astronauts. *NASA*

The ISS astronauts have placed Luca Parmitano's spacesuit under pressure again. Once again, liquid is visible inside the helmet. *NASA-TV*

Astronaut Chris Cassidy points to the ventilation opening through which the liquid was able to get inside the helmet. *NASA-TV*

No one in Houston was aware that Parmitano was squeezing the hand of his colleague Cassidy tightly at that moment. Among astronauts it was the signal for "everything's OK." Cassidy reported to ground control: "He appears to be all right. He doesn't look good, but he's OK."

After this reassuring report, the astronauts in the ISS carried out the pressure equalization with the usual staged increases in pressure. After a seemingly endless nine minutes, they were finally able to open the inner hatch without danger. Chris Cassidy and Luca Parmitano had been outside the space station for one hour and thirty-two minutes.

The four waiting astronauts hurried to help Parmitano get his helmet off. They handed him handkerchiefs, with which he dried his nose, mouth, and ears. Luca Parmitano was finally able to breathe freely again. He was safe. His astronaut colleagues helped the Italian out of his suit. They estimated that they drained 1.05 to 1.6 quarts of water from his helmet.

As usual in such cases, after this incident, which was assigned the highest level of danger, an investigating committee was set up. The committee began working on July 22,

2013. It was able to draw on extensive photographic and recorded material, which had been made in varying degrees of quality during the extravehicular activity. The committee was also able to interview everyone involved in the incident, whether in ground control, in space, or in the affected suppliers.

Two questions dominated the investigation. Foremost was the question of the origin of the water in the astronaut's helmet. The answer as to why it had taken twenty-three long minutes before Cassidy and Parmitano had been ordered to terminate their EVA after the first indication of water in the latter's helmet seemed equally important.

The first of the two questions was answered quickly. Contamination had prevented a tight seal in a connection between the spacesuit's water separator and its cooling lines. Cooling water had made its way into the helmet through this leaky spot. The origin of the contamination could not be determined.

But how could it have happened that ground control took so long to react correctly to Parmitano's indications and end the EVA prematurely? One reason was surely due to the fact that the controller had failed to react appropriately to the presence of water in Parmitano's helmet after the end of the previous EVA. Instead of just ordering the replacement of the drinking bag, a detailed analysis of the incident should have been carried out, especially with respect to a possible danger to the astronaut wearing the spacesuit. The controller was thus responsible for the incident.

The members of the committee found it equally disturbing that Karina Eversley, the controller responsible for monitoring the extravehicular activity, and her support staff had obviously not been aware at any point that the origin of the water in the suit could have been the cooling system, nor had they realized how water might make its way into the helmet. Had they considered the possibility that the water came from the cooling system and not the drinking bag, it should have immediately become clear to them that much more water had to be expected from that source. Eversley might then have reached the correct conclusion and immediately ordered the astronauts back to the space station.

Consequently, one of the most important recommendations addressed to NASA by the investigating committee called for the agency to make its responsible staff more familiar with the technology entrusted to and used by them and with the procedures then in place. The staff was to be immediately made aware of any changes to the hardware. Dealing with emergency situations was to be practiced more frequently in exercises and simulator runs.

Urgent warnings to investigate every unusual situation and take the necessary measures to prevent a repetition followed this incident, just as they had after the *Challenger* and *Columbia* disasters. Not until the staff was certain as to what, how, and in particular why everything had happened could everything return to the usual routine. For the incident of July 9, this meant no more extravehicular activities until all remaining questions about the incident had been answered. Only then could there be a return to normal operations, provided there can and should be anything like normality in space travel.

As an immediate measure, a sort of snorkel was installed in the spacesuits for extravehicular activities. Approximately 19 inches long,

A NASA employee with a model of the snorkel. *NASA*

this tube extended from the helmet into the air-filled spacesuit to about hip level. Should the astronaut fear that he might inhale some sort of liquid or even a solid with the breathing air in the helmet, he can use the snorkel to inhale clean, uncontaminated air. Astronauts on the International Space Station fabricated the first examples of such snorkels from materials available on the space station, so that they would not have to wait for modified spacesuits to arrive from earth.

This system of course served to deal only with an emergency situation involving a massive influx of water. To neutralize small quantities of water, a sort of pad was installed inside the helmet in the neck area. The pad can ab-sorb and store about 2.5 to 3.3 cups of liquid, much like a baby diaper. In addition to absorbing and storing liquids, the pad also makes it possible for the astronaut to feel moisture by touching the pad with the back of his head, providing an early warning of possible danger.

On August 20, Luca Parmitano summarized his views on space travel in general and extra-vehicular activities in particular in his blog: "Space is a harsh, inhospitable place and we are explorers, not settlers. The ability of our engineers and the technology around us make things appear simple, even though they are not. And perhaps sometimes we forget that." Then Parmitano went on: "But we had better not forget that."

THE CRASH OF THE SPACEPLANE SPACESHIPTWO

In the early days of space travel, national institutions held a monopoly on spaceflight. Only national states had the necessary financial and personnel resources for such undertakings. Logically during this time, space vehicles were also developed only by state agencies and were launched by federally owned booster rockets. The crews of manned spacecraft were, as a rule, test pilots with a military background.

At the beginning of the 1960s, however, the United States passed a law allowing private companies to operate communications satellites. These continued to be launched by government-owned booster rockets, however.

Beginning in 1980, Europe came to the fore in the privatization of space travel. As of that year the commercial operation Arianespace began marketing the various rockets of the Ariane series. In the United States it initially remained political strategy to maintain the monopoly for space launches in government hands. This did not change until 1984, when President Ronald Reagan signed a law that permitted American industry to build rockets and operate launch sites at its own expense and responsibility. President George W. Bush later relaxed the stipulations even further by empowering the NASA space authority to approve requests from private companies and institutions seeking opportunities to launch payloads into orbit.

In 2006, the United States finally announced that it intended to employ private companies to transport cargoes into space; for example, to supply the International Space Station. This

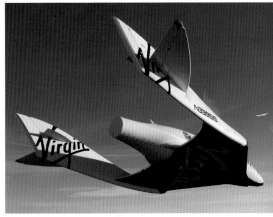

SpaceShipTwo. *Virgin Galactic/Mark Greenberg*

became a reality for the first time in May 2012, when a Dragon spacecraft of the American spaceflight company SpaceX delivered supplies to the ISS.

The company Scaled Composites, a high-tech firm founded by American aviation and space engineer Burt Rutan, has been in business since 1982. In 2007, the company, which is based in Mojave, California, acquired Northrop Grumman, an aviation and armaments company.

During its existence, Scaled Composites has developed various aircraft prototypes and components for aviation and spaceflight, but also small booster rockets, some under contract and some for its own marketing. One of its best-known developments is the balloon cabin, with which several unsuccessful attempts were made to circle the earth by balloon in the early 1990s. The Virgin Atlantic Global Flyer by Scaled Composites did, however, succeed in circling the earth. In 2005, American billionaire Steve Fossett became the first person to fly a jet-powered aircraft nonstop around the world without refueling.

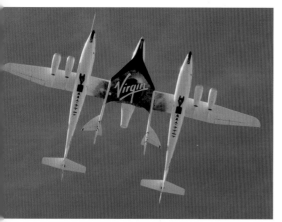

WhiteKnightTwo with SpaceShipTwo, seen from below. *Wikipedia/Jeff Foust*

The Mojave Air and Space Port. *Wikipedia/Ian Kluft*

The Tier One project was the company's first commercial space project. In 2004, the system, made up of two components, succeeded in becoming the first private space vehicle to reach the edge of space, at 70 miles above the earth and beyond. A special aircraft called the White Knight carried a space glider to an altitude of 9 miles, where the vehicle, dubbed SpaceShipOne, was released. It ignited its rocket engine and entered a steep climb toward space. SpaceShipOne accommodated two pilots and two passengers.

The more powerful carrier aircraft WhiteKnightTwo was developed for the follow-up Tier 1b project. It is 78 feet long with a wingspan of 141 feet and carries a payload of 37,478 pounds. The four-jet twin-boom aircraft is designed and built to transport the also newly developed SpaceShipTwo, but it can also carry other payloads, such as small rockets for launching satellites.

SpaceShipTwo was developed and built as the successor to SpaceShipOne by the Spaceship Company, a joint venture of the Virgin Group and Scaled Composites. Since 2012, the Spaceship Company has been a 100 percent subsidiary of Virgin Galactic, owned by British businessman Richard Branson. He had developed the plan to begin carrying out commercial suborbital flights with the space aircraft in 2015, and to use the machine for space tourism.

With a length of 60 feet, SpaceShipTwo was more than 30 feet longer than its predecessor. Its wingspan was 27 feet, and the spaceplane had a height of 15 feet. The machine was designed to reach speeds of up to 2,600 miles per hour. It seated a crew of two and a maximum of six passengers.

The combination of WhiteKnightTwo and SpaceShipTwo took to the air for the first time from the Mojave Air Space Port in California on March 23, 2010. During this first flight the two aircraft remained joined together from takeoff to landing. The first free flight by SpaceShipTwo took place on October 10 of the same year, when, after it was released by the carrier aircraft, the spaceplane glided back to the mission's starting point. Later in the test program the so-called feathering mode, in which the spaceplane alters its wing geometry, fundamentally changing its flight characteristics, was tested successfully on May 4, 2011. After

Experts of the National Transportation Safety Board at the accident site.

sixteen unpowered flights, the first rocket-powered flight finally took place on April 29, 2013.

Fate struck on October 31, 2014, during the twenty-second flight. As it had done several times before, the WhiteKnightTwo cargo aircraft carried the spaceplane to an altitude of 49,200 feet. The pilot of SpaceShipTwo on that day was forty-three-year-old aviation engineer Peter Siebold. The copilot was Michael Alsbury, thirty-nine years old and the father of two. Both men were considered experienced test pilots.

After its release by the cargo aircraft over the Koehn Lake dry lakebed in the Mojave Desert, the pilots ignited the rocket engine. With rapidly increasing speed, SpaceShipTwo began climbing almost vertically toward space. The flight plan called for the feathering system to be manually unlocked by the copilot after reaching Mach 1.4. This initial measure was supposed to make the feathering system ready for rapid deployment at any time, as needed. This was done to avoid mistakenly initiating the return from space without this vital system.

As prescribed, the copilot advised his pilot when they achieved a speed of Mach 0.8. In the cockpit voice and video recordings, the statement "unlock" can be heard seconds later.

Immediately afterward it can be seen that the copilot in fact operates the feathering system's safety lever. In doing so he had clearly shut down the safety system well before the scheduled time. This should not have happened until the spaceplane had exceeded the speed of sound, since experience had shown that heavy vibrations and extreme loads on the spaceplane's structure had to be expected in the transonic range.

The growing vibrations in the seconds that followed reached and finally exceeded the mechanical load limits of the unlocked feathering system. The mechanism failed and the aerodynamic forces in the still relatively thick atmosphere unintentionally activated the feathering mode and caused the wings to fold uncommanded. Altitude at that moment was 59,000 feet. The spaceplane's structure could not stand up to the sudden change in drag. At 10:17:32 Pacific daylight saving time, SpaceShipTwo broke up. Just thirteen seconds had passed since ignition of the rocket engine. Wreckage rained down on an area of more than 3 square miles. It was only because the spaceplane was operating over the unpopulated desert that no one on the ground was injured or killed.

The crew was not as lucky. Copilot Michael Alsbury was found dead in the wreckage, still strapped to his seat. When the spaceplane broke up, pilot Peter Siebold was thrown clear in his ejection seat. Miraculously, he received only a minor shoulder injury.

The time it took for an unprotected human to lose consciousness above 49,000 feet in the thin, cold air due to lack of oxygen is estimated to be nine to twelve seconds. In an explosive pressure loss, such as that which happened in the breakup of SpaceShipTwo, this time is shortened by more than half. Peter Siebold must have fallen unconscious for many thousands of feet before regaining consciousness. He later said that he eventually came to still strapped to his ejection seat in free fall. He managed to free himself from the seat at a height of 16,000 feet above the ground. His parachute then deployed automatically.

The National Transportation Safety Board, or NTSB, the agency responsible for investigating aircraft accidents, began its investigation immediately after the crash. The investigators found that copilot Michael Alsbury had released the safety for SpaceShipTwo's feathering system at Mach 0.92 instead of Mach 1.4 as planned. Investigation into why this had happened soon brought to light the fact that the pilots of SpaceShipTwo had previously complained that it was extremely difficult to unlock the feathering system's safety mechanism at the right time. During the phase of rocket-propelled ascent, the powerful g-forces limited the crew's actions, and they were overtaxed by many tasks that had to be accomplished simultaneously. From these statements the NTSB investigators determined that it was

The Technology of SpaceShipTwo

SpaceShipTwo is a manned aircraft for suborbital flights to altitudes up to about 60 miles, and thus the edge of space. Also called a spaceplane, the two vehicles were developed by the Spaceship Company of California, a subsidiary of Virgin Galactic. Commercial flights with paying space tourists are planned.

SpaceShipTwo is an air-launched space vehicle. It is carried by an aircraft called WhiteKnightTwo to an altitude of about 49,000 feet. After it is released, it climbs under its own power to an altitude of up to 360,000 feet. From there, unpowered like a glider, it glides back to its departure airfield and lands on a normal runway.

The spacecraft is 60 feet long and has a wingspan of 27 feet. It is 15 feet high and its maximum takeoff weight is 21,473 pounds. It is almost completely constructed of carbon fiber composite material. It has a crew of two. The spaceplane can also accommodate up to six passengers, who can experience weightlessness during the four- to five-minute-long unpowered phase of flight. The passengers can also look out at the blackness of space through portholes and see the curvature of the earth. The entire flight from release by the mother ship until landing lasts a maximum of one hour.

SpaceShipTwo is powered by a hybrid rocket engine that combines solid fuel with a liquid oxidizer. This type of power plant combines a simple, low-cost design with relatively high operating reliability. The engine is capable of accelerating the spaceplane to 2,610 miles per hour in seventy seconds. After the engine is switched off, SpaceShipTwo flies like a rifle bullet, propelled by its own momentum on a parabolic trajectory to its service ceiling, and from there it glides back to earth.

SpaceShipTwo uses a special technology, the feathered reentry system, for its return to earth. This system combines the advantages of both of the reentry procedures so far used in spaceflight—capsules landing by parachute and a controlled glide by winged space vehicles such as the space shuttle.

The wings of SpaceShipTwo are sized to provide sufficient lift to allow the spaceplane to climb to its service ceiling after its horizontal release by the mother ship. At the apex of its parabolic trajectory, the rear parts of the wings are then turned upward, up to 60 degrees. This is called feathering mode.

In feathering mode the spaceplane flies like a shuttlecock, which always assumes the same stable attitude no matter how it is thrown. It remains horizontal in the air, with its angled wings at right angles to the air flow. As a result, it develops a great deal of lift, and heat buildup caused by friction remains relatively moderate. Expensive and sensitive heat shields like those used by space capsules or the space shuttle are therefore not required. The thermal conductivity of the carbon-fiber fuselage and wings is sufficient to withstand the friction heat of reentry. Only the leading edges of the wings and the elevator and nose of the spaceplane require special materials to protect them against excessive heating.

When the vehicle's rate of descent drops and the atmosphere is again thick enough for gliding flight, the wings are folded back to their original position, and SpaceShipTwo

again becomes a conventional aircraft with known flight characteristics.

The feathering mode was first tested on May 4, 2011. In October 2014, however, pilot error during a test flight resulted in a malfunction of the system and an accident that claimed one life.

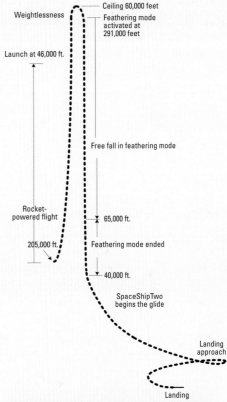

Typical profile of a SpaceShipTwo flight. *FAA/Woydt*

The WhiteKnightTwo mother ship with SpaceShipTwo. *Virgin Galactic / Mark Greenberg*

possible that the copilot had intentionally unlocked the safety system in order to have more time later for other tasks.

As operator of the spacecraft, Virgin Galactic had also obviously failed to brief its pilots on the danger of unlocking the feathering system before exceeding the speed of sound, to say nothing of installing any sort of blocking device. During training in the simulator, crews had been given no or at least insufficient information about the dangers of unlocking the system too soon.

Richard Branson, the founder and owner of Virgin Galactic, used the accident as an opportunity to rethink his involvement in commercial spaceflight. In a blog entry at the turn of the year 2014–15, he posed the question "whether in fact it was right to be backing the development of something that could result in such tragic circumstances." In the same posting, however, Branson wrote that he had been convinced by developers, designers, the pilots, and many other staff that "truly opening space and making it accessible and safe is of vital importance to all our futures."

Virgin Galactic continues its involvement in the field of private commercial spaceflight. The successor to the first SpaceShipTwo, which had crashed, was unveiled to the public in February 2016. The first gliding test flight was carried out on December 3, 2016.